GAME THEORY MASTERY

The Practical Guide to Master Every Decision, Anticipate Others' Moves, and Ensure Success in Work and Life | Become the Leader Who Anticipates Every Move Before It Happens

Strategic Thought Publications

© Copyright 2024 Strategic Thought Publications

All rights reserved

GAME THEORY MASTERY

© Copyright 2024 Strategic Thought Publications All rights reserved.

Written by Strategic Thought Publications, First Edition

No parts of the book can be reproduced in any form without permission from the author.

Copyright Notice and Limited Liability

This notice is legally binding according to the Committee of Publishers Association and the American Bar Association within the United States. Other regions may apply their own legal standards. Any reproduction, transmission, or copying of the material in this work without the written permission of the copyright holder will be considered a copyright violation as per the legislation in effect at the time of publishing and thereafter. Any derivative works from this material may also be claimed by the copyright holder. This book is intended for personal use only. The information within this document is provided for educational and entertainment purposes, and no warranties of any kind are expressed or implied. Readers acknowledge that the author is not providing professional or other types of advice. Although the author has made every effort to provide accurate and current information, no guarantees are made regarding its accuracy or validity, as the author does not claim to be an expert on this topic. Readers are encouraged to conduct their own research and consult with experts as needed to ensure the quality and accuracy of the material presented. By reading this document, readers agree that the author is not liable for any losses, direct or indirect, resulting from the use of the information contained within, including but not limited to errors, omissions, or inaccuracies.

TABLE OF CONTENTS

TABLE OF CONTENTS 5
INTRODUCTION 9
PART 1 12
FOUNDATIONS OF GAME THEORY 12
CHAPTER ONE 13
WHAT IS GAME THEORY? 13
WHAT IS GAME THEORY? 13
HOW GAME THEORY WORKS 14
KEY CONCEPTS 15
CHAPTER TWO 19
CORE CONCEPTS AND TERMINOLOGIES 19
THE NASH EQUILIBRIUM 19
TYPES OF GAME THEORY 20
ELEMENTS OF GAME THEORY 22
EXAMPLES OF GAME THEORY IN REAL LIFE 24
TYPES OF GAME THEORY STRATEGIES 28
CHAPTER THREE 30
MATHEMATICAL FOUNDATIONS 30
GAME THEORY AND MATHEMATICS 30
DECISION TREES AND PAYOFF MATRICES 32
GRAPHS AND EQUATIONS 34
PART 2 37
GAME THEORY IN ACTION: REAL-LIFE APPLICATIONS 37
CHAPTER FOUR 38
GAME THEORY IN BUSINESS 38
NEGOTIATION STRATEGIES 39
COMPETITION AND MARKET STRATEGY 42
CASE STUDY 45

CHAPTER FIVE 49
GAME THEORY IN POLITICS AND INTERNATIONAL RELATIONS 49
STRATEGIC DECISION-MAKING IN POLITICS 49
THE CUBAN MISSILE CRISIS 52
CONFLICT RESOLUTION 55

CHAPTER SIX 59
GAME THEORY IN INTERPERSONAL RELATIONSHIPS 59
RELATIONSHIP DYNAMICS 60
TRUST AND COOPERATION 63
SOLVING COMMON RELATIONSHIP DILEMMAS WITH GAME THEORY 66

PART 3 71
ADVANCED APPLICATIONS AND TECHNIQUES 71

CHAPTER SEVEN 72
EMOTIONAL GAME THEORY 72
BEHAVIORAL GAME THEORY 72
TRUST, RISK, AND COOPERATION 76
CASE STUDY 79

CHAPTER EIGHT 83
GAME THEORY IN EMERGING FIELDS 83
TECHNOLOGY AND INNOVATION 83
MODERN APPLICATIONS 86
ETHICAL CONSIDERATIONS 89

CHAPTER NINE 93
GAME THEORY'S CROSS-DISCIPLINARY IMPACT 93
GAME THEORY AND EVOLUTIONARY BIOLOGY 93
PSYCHOLOGY AND COGNITIVE SCIENCE 96
ECONOMICS AND POLITICAL SCIENCE 100

PART 4 103
PRACTICAL APPLICATIONS AND CASE STUDIES 103

CHAPTER TEN 104
IMPROVING NEGOTIATION SKILLS 104
NEGOTIATIONS IN BOTH PROFESSIONAL AND PERSONAL CONTEXTS.. 108
APPLICATION EXERCISE.. 110
CHAPTER ELEVEN 114
REAL-WORLD CASE STUDIES 114
FAMOUS BUSINESS DISPUTES.. 114
HISTORICAL EVENTS 117
KEY LESSONS FROM THE CASE STUDIES 119
CHAPTER TWELVE 122
IMMEDIATE APPLICATIONS 122
APPLYING GAME THEORY IN DAILY LIFE 122
QUICK EXERCISES.. 125
CUSTOMIZABLE TOOLKIT 126
CONCLUSION 131
EXCLUSIVE CONTENTS: 134
REFERENCES 137

INTRODUCTION

You and your boss are discussing a raise in a meeting. The stakes are high, yet you know you deserve it since you've worked hard. You can't afford to undersell yourself, but you don't want to appear overly demanding. Or, let's imagine you have to make a difficult decision in your personal life, like moving to a new city for work or staying closer to friends and family. These circumstances might seem overwhelming because the results are unpredictable, and the ramifications are significant. But what if you had a secret weapon – a framework that gives you confidence while you made these decisions?

Welcome to *Game Theory Mastery*, the ultimate resource for mastering the art of decision-making. Whether you're trying to make better decisions in your daily life, managing complicated interpersonal interactions, or navigating the complexity of the business world, this book is meant to be your strategic toolset.

Although game theory may seem complicated, its fundamental goal is understanding how individuals make decisions in socially interactive contexts. Game theory allows you to think strategically and predict outcomes in various situations, such as deal negotiation, market competition, or even choosing where to have dinner with friends.

Interestingly, game theory isn't useful solely to economists or scholars. Anyone who wants to make better decisions should use it. It's about planning your actions based on your anticipation of what other people might do. And whether you're at work, home, or wherever in between, you can tip the odds in your favor when you do this effectively.

In this book, even if you've never heard of game theory, we'll begin by understanding the basics in a simple and smart way. We will examine the basic ideas – such as the prisoner's dilemma, Nash equilibrium, and zero-sum games – that will form the basis for the following contents.

And we won't stop there. This book has a ton of practical applications that you can use immediately. Once you grasp the ideas, we'll explore real-world situations where game theory can be helpful, such as commercial negotiations and interpersonal relationships. We'll go through well-known case studies, such as how game theory is applied in the IT sector today or how it helped resolve the Cuban Missile Crisis. And because learning is most significant when hands-on, there are exercises and activities to apply what you've learned immediately.

Starting with the fundamentals and progressing to more advanced techniques, this book will cover it all. You can learn a lot, regardless of your level of experience in strategic thinking. We'll make sure you know how to use game theory in real life and that you understand it as well.

Throughout the book, we'll also explore the ethical side of game theory – how to use these powerful tools responsibly. After all, making wise decisions is more than just coming out on top; it's also about coming up with fair and balanced solutions in life and business.

By the end of this book, you will have your customized game theory toolkit. You can utilize exercises, strategy charts, and templates anytime you need to make a crucial decision. No matter your obstacles, you'll learn to think strategically, anticipate other people's actions, and choose the best path forward.

Your life's journey is shaped by every choice you make. With the techniques and tools this book will teach you, you will be more capable at navigating the challenges. You'll become more confident in your decision-making and strategic thinking and, eventually, succeed in everything you set out to accomplish.

If improving your decision-making, professional standing, and interpersonal relationships is your goal, this book will show you the way. Prepare to change how you perceive decisions and start becoming an expert today. Let's get started!

PART 1

FOUNDATIONS OF GAME THEORY

CHAPTER ONE
WHAT IS GAME THEORY?

Game theory is a mathematical theory with a vast array of applications. Applied in fields as diverse as business, economics, psychology, and even combat, it's essentially the science of strategy. Whether they are potential customers, competitors, or government agencies, it's important to recognize that our plans and activities depend highly on their tactics and actions.

But what is precisely game theory? More importantly, how does it relate to your business in general? Here we'll examine some of the basics of game theory and how it might help you make effective business decisions.

WHAT IS GAME THEORY?

Game theory can be defined as the study of how and why people and other entities – referred to as 'players' – make decisions in various contexts. It's a theoretical framework that facilitates the development of social scenarios between competing participants.

The economist Oskar Morgenstern and mathematician John von Neumann of Princeton University collaborated to develop game theory in the middle of the 1940s. Game theory seeks to determine the interdependent actions that players take in various 'games' to ensure the best results for themselves using logic and mathematics. The games in question may be anything from a duopoly in business to a game of chess.

Game theory can be thought of as the science of strategy, or at least the best way for independent, competing actors to make decisions in a strategic environment.

A wide range of circumstances are mapped out, and their most likely outcomes are predicted using game theory. For example, businesses can use it to decide how to handle litigation, decide whether to acquire another company and set prices. The interdependence of player actions and how they influence the decisions of other players are also examined by game theory, which considers the desired results of various players, regardless of the type of game.

These results could be:

- Positive sum, which benefits both them and the other players.
- Negative total, detrimental to all players.
- Zero-sum, beneficial for a single player at the eventual cost of others.

One example of a zero-sum game is warfare, where the only possible result is the defeat of the opposing player. Another example is when a company seeks to drive its competitors out of business.

HOW GAME THEORY WORKS

The purpose of game theory is to clarify the strategic choices made by two or more participants in scenarios where the rules and outcomes are already established. Game theory can be used to help identify the most likely outcomes when there are two or more players involved, and there are known payouts or measurable outcomes.

The game, which is an interactive scenario with rational players, is the main subject of game theory. The fundamental principle of game theory is that a player's result is influenced by the strategy employed by their opponent.

The game presents identities, inclinations, and strategies available to the players, together with how these strategies impact the result. There may be more conditions or requirements depending on the model.

Applications of game theory are numerous and include business, politics, economics, psychology, and evolutionary biology. Despite its many advances, game theory is still relatively young and developing.

KEY CONCEPTS

Let's first dissect a few important concepts. Gaining an understanding of these fundamentals will enable you to observe how game theory applies to actual scenarios such as negotiations, economic transactions, and even interpersonal relationships. You'll start noticing strategies and outcomes everywhere once you understand the basics of game theory, which are essentially just about using forward thinking to make smarter decisions.

Players: who's involved?

As previously mentioned, in game theory the individuals or groups engaged in a decision-making process are referred to as the players. These players could be businesses trying to close a contract, people trying to resolve a conflict, or even nations choosing how to implement trade laws. The key takeaway from this concept is that any choice a player makes will affect the other players. Decisions are never made in a void; the actions of others matter.

Therefore, identifying the players in the game is the first stage. Knowing who is involved lets you develop better strategies, whether it's you and a competition or you and a possible business partner.

Strategies: what are the options?

Strategies are the various courses of action that every player can choose from. It's all about deciding what move you're going to make, knowing that the other players are thinking the same thing. However, the interesting part of game theory comes when you realize that you're not just picking a strategy on its own; you're also trying to predict the actions of your opponents.

Imagine you're negotiating a deal. You may choose to reply with non-monetary benefits, accept the first offer, or demand additional money. The opposing party is deciding whether to make a stronger offer, hold out, or provide benefits to sweeten the deal with perks. Your strategy will vary based on your expectations of what the other players will do. Similar to chess, you have to continuously plan out several steps ahead of time.

Payoffs: what's the reward?

Payoffs are the results or benefits that stem from a specific approach. In the corporate world, this can mean securing a new contract or finalizing a deal. In real life, it could be as simple as coming to an agreement with a friend. The objective is to select the approach that yields the best outcome for the given circumstances, as each one has a different payout.

The catch is that the payout you receive isn't solely determined by your choices; it also depends on what the other players decide to do. Thus, learning how to optimize your payout while taking into account what others might do is a necessary component of mastering game theory.

Games: the overall scenario

According to game theory, a 'game' is every scenario in which the players' choices influence each other's results. Some games are cooperative, such as when two corporate partners work together to close a contract, while others are competitive, such as when two corporations compete for market share. Every game involves a group of players, each with their own set of tactics, and the way the game unfolds depends on the decisions they make.

The classic example is the prisoner's dilemma. Unaware of each other's decisions, two players must determine whether to work together or turn against one another. Cooperation yields the best results, yet acting out of self-interest can have negative effects on everyone. It's surprising how often these kinds of situations arise, and game theory can help you determine the optimal course of action. Afterwards, it will be analyzed well.

Details: what do we know?

In a game, details are the information available to you at any particular moment. These could be the players engaged, the game's regulations, or even the possible strategies to be employed. Accurate details are essential in each situation because they give more information for the decisions you make. For instance, in a commercial discussion, the details could be the terms of the agreement, the standing of the opposing corporation, and your personal financial goals. The more information you have, the more effective the strategy you design will be.

But in real life, you often have to make choices based on insufficient knowledge. With game theory, you can think through potential outcomes based on what you already know, which helps you manage those scenarios.

Result: what's the outcome?

The result is the conclusion that players reach after completing their decisions. This could be the choice to leave the table, a business arrangement, or a partnership. The outcome is determined by the strategies selected by each player, and often, it's not just about winning or losing but about finding a balance that works for everyone.

In game theory, the outcome is often explained in terms of payoffs. Did you achieve your goals? Did the other players achieve their goals? The best results typically occur when all players understand the rules of the game and make choices that benefit all parties.

Stability: finding the equilibrium

When every player has chosen their course of action, and there is no reason for them to change, the situation is said to be in a state of stability. This is called 'equilibrium'. In other words, you've arrived at a point in which, considering the decisions taken by the players, everyone is receiving the best outcome imaginable.

A famous example of this is the Nash equilibrium, in which each player's strategy is the best response to the other players' strategies. The game approaches stability because there is no longer any gain for anyone to change their strategy. In a real-world commercial negotiation, equilibrium may be attained when both parties agree on terms that are beneficial to all.

CHAPTER TWO
CORE CONCEPTS AND TERMINOLOGIES

Game theory's primary goal is to define precise guidelines and results that make sense of the strategic moves made by two or more players in a particular situation. Game theory can be used to determine the most likely outcomes in any scenario involving two or more parties that have a sizable payout or quantifiable outcomes.

The game (which is, as previously stated, an interactive setting with rational players) is the primary subject of game theory. The payout (or outcome) for one player depends on the course of action chosen by other players.

Before the game starts, a certain number of requirements or conditions about the parties involved are typically required. These include the identities, preferences, and potential effects of the strategies accessible to the player (parties involved), as well as how they may affect the final result. Every decision and action made by the parties affects the result. It's well-accepted that players want to maximize their payoffs and try to act rationally.

THE NASH EQUILIBRIUM

Nash equilibrium – named after John Nash – is a result that, if attained, implies that no participant can increase payoff by making decisions on their own. It can also be considered a 'no regrets' outcome, in the sense that the player will not look back on a decision they have taken, even after taking the implications into account.

The Nash equilibrium is typically attained gradually. Nevertheless, once reached, it will not be altered. While examining the effects of taking unilateral action in such a situation, ask yourself: "Would it even make sense to take action?" The answer is typically no, and that's why the Nash equilibrium result is referred to as 'no regrets.'

Generally speaking, a game can have multiple equilibriums. However, this typically occurs in games where players face more complex decisions than in the example provided here. In games that are repeated over time, one of these multiple equilibria tends to emerge as players adjust their strategies through trial and error.

When two businesses are determining the prices of highly interchangeable goods, such as plane tickets or soft drinks, they frequently find themselves in the dilemma of making multiple choices over time before reaching an equilibrium.

The Nash equilibrium indicates that, as long as they stick to their plans, there will be no gain from changing their actions. But it does not necessarily imply that the best course of action is taken.

To put it simply, the Nash equilibrium can be explained by the following hypothetical scenario: two friends want to share a piece of cake. However, they are unable to agree on what cake. A Nash equilibrium occurs when two people split a cake if they would still prefer to share it rather than purchase one for themselves.

TYPES OF GAME THEORY

Different kinds of game theory arise in a range of topics and areas. Here is a list explaining them, and their possible results.

Cooperative and non-cooperative games

This is the most common type of game theory. When only the payoffs are known, cooperative groups or coalitions engage according to cooperative game theory. This kind of game examines how groups form and distribute rewards among themselves, as opposed to just a two-player game.

On the other hand, the study of non-cooperative game theory looks at the methods used by rational economic agents to achieve their own objectives. The most common kind of non-cooperative game is the strategic game, in which the only elements mentioned are the potential strategies and the outcomes of combining particular options. Rock, paper, scissors is a simple example of a non-cooperative game played in real life.

Zero-sum and non-zero-sum games

When multiple participants actively compete with one another to accomplish the same goal, the game is said to be zero-sum. This means that for every winner, there are losers. This also means that the overall net advantage that was won or lost is equal. In many sporting events, one team wins and another team loses; these contests are therefore zero-sum.

A game where everyone has the simultaneous possibility of winning or losing is known as a non-zero-sum game. Consider business partnerships that are advantageous to both sides and add value. Rather than competing and attempting to win at the expense of the other, both parties benefit from the result.

Investing and stock trading are sometimes seen as zero-sum games. In the end, a market participant purchases a stock, and another trader sells it for the same amount. Nonetheless, since different investors have different risk tolerances and investing objectives, transactions could be advantageous to both sides.

Simultaneous move and sequential move games

Real-world scenarios where two players must make decisions simultaneously with their opponent are common. In the market, competitors are putting comparable plans for product development, marketing, and operations together.

Delays in decision-making might occur when one party wants to watch the other's moves before making their own. This usually occurs in negotiations when one player presents a list of requests and the other has a certain window of time to respond with a list of demands of their own.

Single-shot and multiple-shot games

A single scenario can initiate and conclude a game. Like almost everything in life, the underlying competition starts, progresses, ends, and cannot be reproduced. This is frequently the case for stock traders, who must decide when to join and leave the market based on information that may not be easily changed.

On the other hand, certain games appear to go on forever and they are played a lot. The players in these types of games are usually the same, and all parties are aware of each other's previous behavior. For instance, think of opposing businesses who want to set their own prices for their products. Every time one changes its price, the other might do the same. Regardless of product cycles or sales seasonality, this never-ending competition exists.

ELEMENTS OF GAME THEORY

A group may be faced with a variety of situations, and the decisions its members make will impact the outcome. The members of the group collaborate to examine different combinations and choose which one will produce the best result. This is not a theory that only holds true for team members. It has also had a significant role in how businesses choose locations for their operations. Understanding the many components of the theory is helpful when working on a topic in a group environment. Here are a few possible scenarios.

When there's no perfect decision

Even after examining every option, it's often the case that no decision is perfect. However, based on the information available to them at any particular moment, the decision-makers come to a conclusion. An example of this would be a situation in which a group of people agrees on a course of action based on scarce knowledge of the circumstances.

When making a rational decision

All participants in the process make logical decisions. A person can make decisions free from the impact of emotions or chance by applying rationality. All participants applied rational thinking using game theory.

When individuals communicate and interact

No matter how challenging the process may be, those engaged reach a decision by interacting and using effective communication techniques. Every player's actions have an impact on other players. Having effective communication can help prevent conflicts.

When players aim for personal fulfillment

It's common for players to put their own interests first during this process. Any confrontation has the potential to highlight this even more. Participants attempt to make decisions based on what they know, what they can offer, and what they stand to gain or lose. A player may have greater autonomy within the team process if they act in their own best interests.

EXAMPLES OF GAME THEORY IN REAL LIFE

Game theory is used in many aspects of daily life, particularly in decision-making. Here are a few instances where it is applicable.

Real estate negotiation

Real estate discussions often include the application of game theory. It applies particularly in situations where the buyer is up against other bidders for the property in a multiple-bid scenario. Many choices and decision-makers are involved. A buyer has three options if they find out that there are multiple counterbids on the property: sticking to the initial offer, backing out by declaring that they have no intention of continuing the negotiations with the other parties, or offering a better deal.

The buyer has the option to outbid a competitor to seal the deal if they so want. In this case, the buyer determines the amount over the price to which they are willing to extend their offer. However, there's still a chance that another bid will come in. This can happen if a late bidder enters the auction or if one of the interested parties raises their offer. Even if the buyer is unable to close the agreement, they will have exhausted all means to acquire the property. They calculated their steps using the knowledge available to them at the time.

Cost of products

Businesses use game theory to determine how much they should charge retailers and consumers for their goods. Retailers compete with one another by providing exclusive deals and incentives. They might have eye-catching sales at certain seasons of the year, like Christmas or the summer, to draw clients towards them. They keep an eye on competitors to see what offers they make, then attempt to equal or outbid them. Game theory is being used by customers as well as retailers. The best product at the best price is what the consumer seeks out. The retailer looks for the best pricing offer in order to attract customers.

The prisoner's dilemma

The game theory example that is most widely known is the prisoner's dilemma. Consider a scenario involving two criminals who were caught in the act, but the prosecutors have insufficient evidence to convict them. In order to get a confession, officials remove the prisoners from their isolation cells and question each one of them separately in separate rooms. Communication between inmates is impossible. Authorities usually present four deals:

- If they both confess, they will serve 5 years each in prison.
- If prisoner B confesses but prisoner A does not, prisoner B will serve 0 years, and prisoner A will serve 10 years.
- If prisoner A confesses but prisoner B does not, prisoner A will serve 0 years, and prisoner B will serve 10 years.
- Finally, if neither of them confesses, they will serve 1 year each in prison.

In this situation, saying nothing is the best thing to do. Both are likely to confess and receive an eight-year prison sentence because neither one is aware of the other's strategy. A prisoner's dilemma is a game in which every player will decide what is best for them individually but worst for the group as a whole.

'Tit for tat' is typically considered the best course of action in a prisoner's dilemma. The idea of 'tit for tat' was introduced by Anatol Rapoport. In an iterated prisoner's dilemma, each player takes an action that corresponds to their opponent's previous action towards them. An example of this would be a player who, if provoked, might retaliate, but otherwise might cooperate.

The following table represents a variation of the Prisoner's Dilemma, illustrating the possible outcomes of choices made by two participants (let's call them Prisoner A and Prisoner B), with one selecting a row and the other a column. Each participant has two options: **Confess** (Row 1 for Prisoner A and Column 1 for Prisoner B) or **Deny** (Row 2 for Prisoner A and Column 2 for Prisoner B). The numbers in each cell represent the prison time each would serve based on their combined choices:

| | | COLUMN'S SELECTION ||
		C1	C2
ROW'S PICK	R1	5, 5	10, 0
	R2	0, 10	1, 1

This table helps to visualize the dilemma: while confessing may seem safer individually, if both deny, they collectively achieve the least total prison time.

Dictator's game

In this simple game, player A must decide how to split a cash prize with player B; however, player B cannot sway Player A's decision. This offers some interesting insights into human behavior, but it's hardly a game theory strategy. Studies show that about half of participants keep the whole amount to themselves, 5% split it equally, and the other 45% give the other person a smaller portion.

A similar game is the ultimatum game: Player A is given a certain amount of money, part of which must be delivered to Player B, who can choose to accept or reject the amount. However, if player B rejects the amount offered, then neither A nor B will get anything. Games like these can teach us important things about kindness and our behavior as humans.

Volunteer's dilemma

In this dilemma, one person must perform an action or duty for the good of the group. The worst-case scenario occurs if no one offers to help.

For example, consider a business where senior management is unaware of widespread accounting fraud. The fraud is known to a few junior employees in the accounting department, but they are afraid to tell higher management for fear of the fraudsters' termination and possible legal action. Moreover, being dubbed a whistleblower may have specific repercussions in the future. However, if no one volunteers, everyone's job could be lost in the event of the company's ultimate collapse due to the massive fraud.

The centipede game

In this extensive-form game, two players alternately take the larger share of a money stockpile that is continuously increasing. Before the game begins, each player knows how many rounds it will have in total.

Due to the setup, if one player passes the pot to the other and they take it in the next round, the reward they receive will be slightly less than if they had taken it this round. However, after an additional pass, the potential payoff grows. This means that although each player has an incentive to take the pot in each round, they would benefit more by waiting.

When one player takes control of the stockpile, the centipede game ends; that player receives the larger share, while the other player is left with the smaller share.

TYPES OF GAME THEORY STRATEGIES

In game theory, players can choose from multiple primary strategies when playing a game. After considering all the information in their possession, each player has to decide how much risk they can take and how far they will go to get the best outcome.

Maximax approach

A Maximax strategy implies that it's all in or all out for the players: they can either win big or have the worst possible outcome.

This could be the case of a start-up business moving its first steps into the market. Its new goods have the potential to increase the company's market capitalization fifty times. But at the same time, the company could go bankrupt due to a poorly executed product launch. A participant is prepared to take a chance in order to achieve the best outcome, even in cases where the worst-case scenario is plausible.

Maximin approach

Using a maximin strategy, a player aims to secure the best possible result among the worst outcomes. This approach helps avoid the riskiest situations. Companies often use this tactic when considering legal disputes. By choosing to settle out of court, they may accept a less favorable result to avoid the risk of a much worse outcome if they went to trial.

The dominant approach

A player employing a dominant strategy acts in a way that optimizes the potential of its own strategy, no matter the actions of the other players. This can imply, in the context of business, that a company decides to expand and join a new market whether or not a competitor has taken the same course of action.

Pure strategy

Pure strategy is essentially a predetermined decision made without consideration of external circumstances or the actions of others, involving the least amount of strategic decision-making.

In the game of rock, paper, scissors, let's examine a scenario in which a player consistently plays the same shape. That player's outcome is known in advance and can take one of two forms: either the same specific shape, or he doesn't participate at all. That's what makes the strategy 'pure'.

Mixed strategy

While choosing which elements or actions to combine requires deliberate thought, a mixed strategy may appear to be the outcome of luck.

Consider the relationship that exists in baseball between a pitcher and a hitter. Pitching in the same manner each time is not permitted. The batter could predict what would happen next if this wasn't the case. The pitcher has to change their approach from pitch to pitch to create an illusion of unpredictability that they hope to exploit.

CHAPTER THREE
MATHEMATICAL FOUNDATIONS

It's easy to get engrossed in the strategic aspects of game theory – the mental struggles, decision-making, and back-and-forth of trying to outwit opponents. However, something a little more tangible lies underneath all of that strategy. But don't worry, you don't have to be an expert in arithmetic to understand how these figures and equations aid in our decision-making. Often, the mathematical aspect involved in game theory merely serves to clarify and illustrate the decisions we encounter daily. Consider this: you're effectively playing a game anytime you have to make decisions about relationships, business, or even simple things like what to eat for lunch. Thanks to math, we can break down the 'rules' of that game, evaluate our decisions, and predict potential results. It's about better understanding the choices before you, not complex mathematics or abstract theories.

In this chapter, we'll look more closely at some basic mathematical concepts underpinning game theory. We'll go over concepts like decision trees, matrices, and probabilities – all of which are tools to help you organize and logically consider your options. If math isn't your strongest subject, don't worry: we'll keep it simple and make sure it's easy to understand. By the end of this chapter, you'll see how math can be a strong ally in making wise, strategic decisions.

GAME THEORY AND MATHEMATICS

Game theory uses mathematics to represent the decisions, risks, and possible results associated with strategic thinking. The purpose of math is to clarify the alternatives, not add complexity. When presented with various possibilities, it can help you make better selections by organizing and visualizing the multiple routes. Matrix algebra, probability, and payoffs are the three main mathematical subfields that make up the framework of game theory. Let's try to explain those briefly.

Matrices

In game theory, a matrix is a way to organize the options for each player in a game. Imagine it as a table or grid, with each row representing a possible move by one player and each column representing a potential response by the other player. Consider the scenario when you are negotiating a deal with a competitor. You have two options: offering a higher or lower price. Your competitor has the same options. You can see how choices result in different consequences by visualizing the possible actions and outcomes in a matrix. It's similar to having a road map of potential future events shown before you.

Probability

Game theory recognizes that there are many unknowns in life. Here's where probability comes into play.

Probability is a tool to help you handle circumstances where results aren't sure. For example, in a negotiation, there may be a 60% chance your competitor will accept your offer and a 40% chance they won't. You can determine your expected reward for each option by accounting for those probabilities, which will help you make more risk-aware decisions. Using probability to your advantage and calculating your possibilities of success (or failure) based on the facts are vital in decision-making.

Payoffs

Finding the solution that works best for you is the primary goal of game theory, and here is where payoffs come into play.

A payoff is the outcome or reward of a particular decision. In business, it could be the money you make; in a personal environment, it could be the fulfillment that comes from coming to a compromise or emerging victorious in a dispute. Every decision in a game or strategic circumstance has a specific payoff. Game theory helps you determine the most rewarding choice by giving these outcomes a value. Depending on the circumstances, the return might occasionally be more than just money; it might also be emotional or reputational.

DECISION TREES AND PAYOFF MATRICES

When presented with a difficult choice, it might be intimidating to consider your options, project possible consequences, and choose the best course of action. Payoff matrices and decision trees are helpful in this situation. By visualizing your options and potential outcomes, these tools help you evaluate even the most difficult decisions.

Decision trees

A decision tree is exactly what it sounds like: it's a branching diagram that shows all of your options and the possible outcomes of each choice. Let's say you have to make a crucial business decision. Should you introduce a new product now or put it off until the following year? You can break this down step-by-step with the use of a decision tree.

The decision you are about to make is like a base from which different options branch out. These branches represent your choices, like whether to wait or start immediately. Each choice leads to different possible outcomes, which are like additional branches stemming from the original ones. If you launch today, the market might not be ready; thus, the product can fail or, on the contrary, have great early sales. If you wait, a competitor might get to market before you do, or you might have more time to refine the product. By the end, it will be easier to understand the pros and cons of each route once you have laid out all of your options and their outcomes.

Decision trees are effective because they make you think through every scenario. Rather than depending on intuition or insufficient data, you can plan and assess the potential outcomes of every choice. This visualization allows you to overcome ambiguity and make decisions more confidently.

Payoff matrices

While decision trees help examine a single choice with several possible outcomes, a reward matrix facilitates side-by-side comparisons of various techniques. Consider a payout matrix as a grid that arranges each player's potential movements in a strategic game. It's beneficial when you're up against someone else, like in a marketing plan competition or business discussion.

Assume that you are debating whether to maintain your current pricing level in your business or cut it. You can compare your plans to those of your competitors using a payoff matrix, and you can then examine the results of each combination of choices. For instance, you might continue to make consistent profits if you maintain high prices. You could gain market share if you cut your pricing while they retain their high ones, but your profit margin would decrease. You can assess the potential impact of each option by reviewing the matrix, which illustrates the payoffs in an easy-to-understand format.

Payoff matrices help visualize action effects, much like decision trees do. However, a reward matrix considers other players' strategies in addition to your own selections. This is particularly useful in competitive settings where you have to predict your opponent's movements, too, and modify your approach accordingly.

How do these tools work together?

The aim of both reward matrices and decision trees is to bring clarity. They assist in organizing thoughts, evaluating all options, and ultimately making better decisions. These methods help break down complex situations into smaller, more manageable parts, whether dealing with multiple outcomes from a single choice or attempting to anticipate the actions of other players. And the best part? If you put them to good use, you'll discover they can help you in ordinary situations and make significant, life-changing decisions.

GRAPHS AND EQUATIONS

Let's explore additional mathematical tools in game theory, like graphs and equations, alongside matrices and other methods we introduced earlier. Don't worry—these concepts are presented in an easy-to-understand way. These tools work together to show relationships, outcomes, and strategies more clearly than words alone can.

Graphs

Graphs are a great visual representation of choices, plans, and results. Assume you are trying to predict how two businesses in the same market will respond to one another's pricing strategies. Rather than trying to monitor everything just by keeping it in mind, a graph can provide you with an accurate visual representation.

In a simple two-player game, we can use a graph to show how each company's pricing decisions affect their outcomes. Imagine one company's pricing choices on the x-axis and the other company's on the y-axis. Each point where the lines meet represents a different outcome based on their combined choices. For example, if both companies cut prices, they may lose profit, but if both keep prices high, their profit margins could stay stronger. This visual helps us quickly see how different choices lead to different results, making it easier to understand the impact of each decision.

This visual approach makes it easier to grasp how each decision interacts with the other and affects both players' and individual choices. You can quickly notice patterns or results that wouldn't be as clear if you were looking at raw numbers or descriptions.

Equations

Although they may seem scary, game theory equations are just a means of figuring out possible results, or payoffs, in a game. Consider the well-known Prisoner's Dilemma scenario. Two of the criminals in this scenario are apprehended and questioned independently. They have two options: they can admit they did it or not. The actions made by both players determine the outcome, and the payoff is the number of years they spend behind bars.

Here is a simple equation that illustrates potential payoffs:

- Should they admit, each faces five years in prison.
- If one admits guilt and the other remains silent, the one who admits guilt serves no more than a year, while the silent person receives ten years.
- If neither speaks, a year each.

Numbers in an equation can be used to show the reward for each scenario. For example:

Payoff (Player 1) = -5 if both confess

Payoff (Player 1) = 0 if player 1 confesses and payer 2 stays silent

You can compute and compare the potential results of each decision using this mathematical approach. You can quickly determine which approach might yield the most significant outcome for each participant by entering different numbers for each scenario.

A more in-depth knowledge of game theory can be gained through graphs and equations. Equations help calculate the possible outcomes, while graphs illustrate the connections between the decisions made by players and the outcomes. Both tools can help you make smarter, more informed decisions in real-world scenarios, whether you're managing a project, negotiating a transaction, or just trying to choose the best course of action for overcoming an obstacle.

After some practice, these tools – which initially seem unpractical – become very helpful in aiding you in seeing the bigger picture in any game or situation involving decision-making. And remember, using them doesn't require being a math expert. Developing structure and clarity is the key to approaching every decision with confidence.

PART 2

GAME THEORY IN ACTION:
REAL-LIFE APPLICATIONS

CHAPTER FOUR
GAME THEORY IN BUSINESS

It's an easy concept to grasp: a business leader uses a variety of factors to inform their managerial decisions and strategy, such as anticipating the actions of competitors and strategizing how to respond to those actions. Game theory offers a mathematical framework to examine the path of action most likely to produce the intended results, assuming accurate projections. Leaders who prepare in this manner can make better decisions regarding pricing, product introductions, target markets, and marketing campaigns. Essentially, "securing a competitive advantage" involves using game theory to achieve desired results.

But when it comes to business decision-making, game theory typically involves more than just a straightforward, binary link between the behaviors of two competitors. In any industry, there are sometimes a lot of competitors in the field. Every move a competitor makes might set off a series of events and reactions that last for a long time. The number of elements increases and makes any accurate long-term result prediction increasingly challenging. Furthermore, a successful business growth plan often depends on a combination of innovation, stability, and occasional disruption. However, estimating how innovation and disruption affect a business also requires risk evaluations and cost-benefit calculations.

By mapping the likelihood of various scenarios, a dynamic game theory approach would also take unforeseen factors into account. The strategic leader will then have several backup plans available. Like chess players, business executives use game theory strategies for diverse situations and anticipate multiple moves ahead. When necessary, they get ready for quick, planned strategy iterations. In turbulent times, this expectation can support business growth while maintaining reactiveness and competitiveness.

NEGOTIATION STRATEGIES

In a chess game, you're trying to figure out how to maximize your benefits while avoiding risks, predict your opponent's next move, and anticipate their plan. Game theory acts as a 'cheat sheet' for this process, guiding your strategic thinking and decision-making. When negotiating a contract, asking for a raise, or closing a business deal, game theory can give you a significant edge by enabling you to understand the bigger picture.

Supplier agreements: when to press for change and when to give in

Assume that you and the provider are discussing a deal. They have their bottom line to defend, but you want the best deal. This is a classic game theory scenario in which each party pursues its goals and tactics.

According to game theory, you and the supplier participate in a game where you each attempt to maximize your profit. Their payoff might be the most significant profit margin, but yours might be the best pricing. Making a reward matrix is one strategy you may use to compare various results:

- If the supplier insists on a higher price, you may accept it in exchange for better service or order priority.
- If you push too hard for a lower price, the supplier may agree but provide less favorable terms, such as slower delivery times or inferior quality.

By imagining these scenarios, you can see situations when a compromise might benefit both sides. For example, you might be able to provide a lower price now in exchange for a longer-term contract, giving the supplier security and enabling you to save money. The secret is to realize that discussions aren't necessarily zero-sum if the appropriate agreement is done.

Salary negotiations: playing your cards right

Imagine haggling over your pay at a new job. As the company tries to control expenses, you seek the best offer. Using game theory can make your approach more intelligent.

In a pay negotiation, you can consider the power dynamics by applying concepts such as the Nash equilibrium. You know that pushing too hard could result in them selecting someone else, and the employer knows they need to give you a fair offer to keep a strong candidate. Game theory aids in identifying the ideal situation where all parties are content.

Here's a strategy: instead of concentrating just on compensation, consider including other payoff factors like bonuses, vacation time, or benefits. For instance, the employer can be open to providing extra benefits that raise your satisfaction level even if they cannot match your wage expectations. You give both sides more interesting opportunities for a mutually agreeable contract when you broaden the negotiation beyond a single figure.

Business transactions: looking past the immediate win

When negotiating an investment or commercial relationship, the stakes can be considerably more significant. In these complicated situations, game theory can help you by enabling you to plan multiple moves ahead of time, much like in a chess game.

Consider yourself negotiating a joint venture with a different business. Iterated games, in which the game is played across several rounds rather than just once, are a crucial concept in game theory. This occurs often in the business world.

Your choice now impacts your relationship with the other party in the long run.

For instance, you can win the contract right away but damage your long-term partnership if you adopt an aggressive attitude and try to control the discussion. However, if you cooperate and show flexibility today, you will establish credibility and pave the way for future conversations that will be more fruitful. Game theory encourages you to consider future payoffs that result from building goodwill and trust in addition to the immediate payoff.

Example: the prisoner's dilemma in negotiation

Again, the prisoner's dilemma aptly demonstrates how this works in negotiations. Let's use it to discuss a pay negotiation:

- Both you (Player 1) and the employer (Player 2) must decide whether to act in your own best interests (you demand an outrageous raise, they present a lowball offer) or cooperate (you ask for a fair raise, they offer one).
- You both have a satisfied outcome if you both work together. They retain a valuable employee, and you receive a respectable raise.
- It is worse for both of you if one of you behaves selfishly. Overly demanding terms run the risk of ending the offer altogether. Both of you lose if you don't cooperate; if they lowball, you might leave the company, and they lose a valuable employee.
- Suppose neither of you cooperates; neither wins. You might continue in an unsatisfactory position, or they might lose you to a competitor.

This makes it more evident that finding common ground rather than going for the selfish option in many discussions is the preferable course of action.

Understanding how your choices affect the other party's and your own results is critical.

Applying game theory in everyday negotiations

Game theory has the advantage of being applicable to everyday conversations, as well as high-stakes trades. The same rules apply whether you're haggling with a contractor, organizing a trip with your significant other, or finding the best bargain on a new car. The goal is to understand the intentions and potential courses of action of each party, then exploit that understanding to shape events in a way that best serves your interests without causing conflicts or leaving any request unfulfilled.

You'll find that you make better decisions and have greater confidence in any negotiation by carefully considering the strategies and potential results. Thus, remember that game theory is on your side the next time you negotiate!

COMPETITION AND MARKET STRATEGY

Every choice you make in a competitive business environment, such as entering a new market or establishing prices, can feel like you're playing a high-stakes game. Both you and your opponents are observing each other's movements. Game theory provides a framework for considering these competitive scenarios and helps formulate winning strategies.

Pricing strategies

Pricing strategy is one of the critical areas in which game theory can be helpful. Assume that you are in charge of a company and must decide on the cost of your products and services. The problem is that your competitors operate under the same conditions, so they might immediately follow suit each time you modify your pricing. How can you be sure that you're not leaving money on the table and not pricing yourself out of the market?

According to game theory, you and your competitors are players in this game, and the game's objective is to maximize your profits. The secret, though, is realizing that your earnings are influenced by what you and your competitors do. This is the application of ideas such as Nash equilibrium. Nash equilibrium in pricing strategy occurs when all participants (you and your competitors) select prices. At this time, assuming that the others maintain their prices unchanged, no one can change their outcome.

For instance, if you and a competitor offer too low prices, you may draw in more business but see a decline in profit margins. You risk losing business to a competitor who offers better if you both charge too much. The Nash equilibrium in this scenario is the pricing point at which you both make enough profit to be pleased, and neither of you is motivated to lower prices more.

The airline sector is a prime illustration of this in the real world. Airline companies often change the price of their tickets in response to actions taken by their competitors. If one lowers its pricing to fill seats, competitor airlines might do the same to keep customers. However, if they continue to undercut one another, they risk a pricing war in which everyone loses. The application of game theory helps airlines identify the ideal price-setting range that prevents a downward pricing spiral.

Market entry

Choosing to enter a new market is another crucial business decision. This is a strategic game because competitors may change their strategies after you join a market. Game theory helps you think about these decisions more strategically, considering if entering a market will pay off or whether it would generate a reaction that can affect your business.

Imagine you're thinking about introducing a new product or growing your business in a new area. The first thing you want to know is how your competitors will respond. The study of game theory helps you prepare for all possibilities. Will they try to evict you by lowering their prices? Will they change their offerings or start vigorously marketing to counter your new product? Payoff matrices help you visualize potential outcomes in situations like these. For instance, you may experience a brief decline in profits if you enter the market and your competitors react aggressively. However, you may be able to establish a significant presence if you join and your competitors remain silent. Using game theory, you can evaluate these situations to see whether the possible benefits outweigh the risks.

Think about the techniques IT giants like Samsung and Apple use to introduce new products. Samsung has to choose how to respond when Apple announces a new iPhone. Is it okay for them to release a competitor product simultaneously? Cut the cost of their current models? In this intense competition, both businesses continuously assess how their actions impact one another. They are aware that making a mistake could cost them market share.

With the aid of game theory, these businesses can anticipate the moves of their competitors and map out possible strategies. For instance, Samsung can conduct a marketing campaign emphasizing its benefits in response to Apple releasing an iPhone with new features, or it might provide a brief discount to entice people to stay with Samsung instead of straying away from them. In addition to competing, the objective is to ensure that their choices do not put both businesses in a losing position.

Striking a balance in competition

The main lesson from game theory in highly competitive contexts is that establishing strategies that lead to a sustainable balance is more critical than just overcoming the competition. Regarding pricing or venturing into new markets, game theory pushes you to consider the consequences of your decisions and actions. The true power is in realizing that making wise decisions that result in long-term success rather than winning every round is what matters most in business.

Competitive business environments are similar to multi-round games in which decisions are made based on previous choices. Sometimes, the most excellent course of action is to achieve stability that helps your business and puts you ahead of the competition rather than taking a daring and risky step.

Thus, next time you decide on your prices or consider growing remember that you're not only playing your own game. You are a participant in a huge multiplayer game. However, with game theory, you are guaranteed to be always one step ahead!

CASE STUDY

Let's examine a real-world business dispute that was settled by applying the concept of game theory, which you can easily use for your business challenges. The OPEC oil pricing dispute is a well-known instance of how game theory has influenced the way oil-producing nations set prices without going into a lose-lose situation.

The OPEC dilemma: a game of cooperation and competition

OPEC (Organization of the Petroleum Exporting Countries) comprises some of the world's largest oil producers. Its member countries have considerable influence over oil prices because they effectively control a sizable percentage of the global oil supply. However, that same authority also raises the possibility of conflict. Every nation seeks to sell as much oil as possible to maximize its earnings; yet, if all nations oversupply the market, prices will fall, and all nations will miss out on potential income.

This is where game theory comes into play. One of the most popular illustrations of game theory, the prisoner's dilemma, is comparable to the OPEC's situation. The optimal solution to this dilemma would be for each nation to secretly boost output while the others stick to their quotas. But if every nation took this action, oil prices would plummet, making things worse for everybody. As a result, competition (self-interest) and cooperation (maintaining stability) must remain balanced.

The role of game theory in resolving the conflict

Due to overproduction, OPEC countries found themselves in a position in which world oil prices declined in the 1980s. Each member nation was enticed to boost oil production to gain a larger portion of the market and earn more money; nevertheless, this resulted in an excess supply, which drove down prices. Everyone was losing money, and the group known as OPEC was on the edge of disintegrating.

Game theory techniques were applied to find a solution. Cooperation and establishing output quotas that all parties could agree upon were crucial in preserving market stability. OPEC countries realized they would all gain if they could achieve a balance where no one had the motive to overproduce, as opposed to each nation behaving individually and selfishly.

They imposed production restrictions, and although it was tempting for individual nations to overproduce, the fear of global production escalation restrained them. It was an ideal illustration of how Nash equilibrium functions in the actual world. The nations concluded that honoring the agreement was the best course of action for each of them in light of what the others were doing. Oil prices recovered due to the production stability, and the emergency ended.

Why does this matter for your business?

An essential business lesson from the OPEC example is that collaboration can occasionally provide more remarkable outcomes than direct competition. Game theory principles help consider the big picture, whether you're negotiating with partners, fixing prices with competitors, or settling internal issues. You can assess what would be the best course of action for you while also taking other people's actions into account.

Picture yourself in a pricing battle with a competitor company in your sector. Should you both continue to lower your pricing to attract more customers, you may find that both company revenues are diminished. However, it's a win-win if you can strike a balance and realize that there is a price point that suits you both.

Game theory teaches you to anticipate other people's responses and realize that your decisions don't happen in a void. The OPEC example demonstrates that sometimes the best course of action isn't to outcompete your competitors but rather to develop a solution that benefits everyone in the long run.

What does the OPEC case teach us?

Consider situations in your company where collaboration could produce better results than competition. Perhaps it's partnering with other businesses to find beneficial prices, negotiating agreements where both parties feel successful, or forging partnerships with suppliers. Game theory applies strategy and logic to determine the optimal course of action in any given economic scenario, not simply doing deals in elite international organizations. Like OPEC, your company will gain from putting long-term stability and cooperation ahead of short-term gains. That is the application of game theory's strength!

CHAPTER FIVE

GAME THEORY IN POLITICS AND INTERNATIONAL RELATIONS

Effective strategy is key to achieving many goals in government, such as creating policies, delivering services, and ensuring public services run smoothly. Strategic thinking helps develop good strategies, which are crucial in managing complex and changing situations. This approach can help inspire better collaboration and resilience when working with others. But how can you develop strategic thinking? And how do you make sense of complex, shifting scenarios?

STRATEGIC DECISION-MAKING IN POLITICS

There are many stakes in politics, and any choice can affect policy, change the course of history, or even transfer power. Unbeknownst to them, politicians often use game theory to make strategic decisions, whether developing policies, establishing coalitions, or running for office. Let's break this down to show you how game theory applies to real-world political situations and is not only a theoretical construct for businesses or academics.

Using game theory in political campaigns

Politicians play a sophisticated game during elections, with voters acting as the 'players' they must win over. Let's say there are two contenders for the same office. They have to choose where to run their campaign, how much money to spend on advertisements, and which policies to prioritize – all while speculating about what their competitor will do.

This is where the importance of game theory comes in. The candidates understand that they cannot achieve everything, therefore they must carefully weigh their alternatives. To make sure they maintain support, should they concentrate on their strongholds or the swing states where the vote is tight? Should they focus on their own positive message or turn against and attack their opponent's? Every decision they make is based on their assessment of what the opposition will do and their prediction of the reaction of the electorate.

In the 2016 United States presidential election, for instance, both campaigns employed game theory to concentrate their resources on crucial areas where the result of the election was unpredictable. They invested a great deal of money in these states because they realized that getting enough Electoral College votes to win the presidency would be more important than gathering the most votes nationally. In order to reply with counterstrategies, both contenders had to predict where their opponent would concentrate.

Forming alliances

Direct competition isn't always the focus of politics. Politicians often join forces with other parties or people in order to improve their chances of succeeding. This is where the coalition-building idea from game theory is useful.

Consider this example: a political party can secure the necessary number of votes by forming an alliance with smaller parties, even if they do not have enough seats in the Parliament to pass legislation or gain a majority. The problem, however, is that each party has different goals and priorities. Game theory helps politicians navigate this difficult issue by evaluating what each side stands to gain or lose and striking a balance where collaboration benefits everyone.

One concrete example comes from countries like Germany or Israel which have several political parties and typically have coalition governments. Parties must form coalitions if none of them received enough votes to win enough seats in the Parliament to rule on their own after elections. Each party must think about how much they can give up to their partners while still pursuing their own goals, in addition to what is best for them. It's a delicate balancing act, much like the prisoner's dilemma in game theory. The alliance may fall apart if one party breaks a deal and betrays the other.

Game theory and policy decisions

Game theory becomes even more important when deciding on policies. Politicians often have to take into account not only the immediate effects of their decisions but also the opinions of other stakeholders of the government, including the public, interest groups, other governments, and other politicians.

A government may be choosing, for instance, whether to apply new taxes on imports. It appears to be a simple approach to safeguard domestic industry at first glance. However, the government needs to predict (based on game theory) how other nations will respond. Will they levy duties in retaliation against exports? How would that impact domestic workers and companies if they do? Game theory is specifically created for this kind of decision-making process, which entails sketching out possible moves and countermoves in a series of exchanges.

A well-known example of game theory applied to international policy is the 1962 Cuban Missile Crisis. During this tense standoff, both the United States and the Soviet Union had nuclear weapons ready for use, putting the world on the brink of war. Game theory helped guide the decisions made by U.S. President John F. Kennedy and Soviet Premier Nikita Khrushchev. Each leader had to carefully anticipate how the other would respond to their actions:

should they make an aggressive move, or try to defuse the tension through negotiation? By using game theory, both leaders could weigh their choices against the potential reactions of the other side, ultimately guiding them toward a peaceful resolution.

Game theory provided a structured way of evaluating possible strategies and outcomes in a situation where each side's actions were highly interdependent. This framework allowed both leaders to consider not just their immediate options, but also the likely reactions and counter-reactions that could escalate into a catastrophic conflict. The final resolution—Khrushchev agreeing to remove missiles from Cuba and Kennedy pledging not to invade—demonstrated the power of strategic thinking in defusing a crisis with global implications.

THE CUBAN MISSILE CRISIS

The Cuban Missile Crisis remains one of the most famous examples of game theory applied to real-world, high-stakes negotiations. For context, in 1962, the Soviet Union's decision to place nuclear missiles in Cuba led to a 13-day standoff that brought the world perilously close to nuclear war. Game theory enabled Kennedy and Khrushchev to assess not only their own goals and strategies but also the potential responses of their opponent. By understanding that each choice carried significant risks and that mutual cooperation was necessary to avoid disaster, they were able to navigate a dangerous situation without triggering a conflict.

Today, political leaders use similar game-theoretic approaches to make strategic decisions, negotiate alliances, and craft policies that consider the actions and reactions of others. Game theory encourages leaders to think broadly, evaluate all possible outcomes, and understand how their decisions impact others, providing a powerful tool for navigating complex, interdependent scenarios.

The setup: the players and their strategies

President John F. Kennedy of the United States and Premier Nikita Khrushchev of the Soviet Union were the players in the Cuban Missile Crisis. The stakes could not have been higher: both nations possessed enough nuclear weapons to completely destroy one another, and both leaders were aware that one mistake could result in full-scale conflict.

After American spy planes found Soviet weapons in Cuba, Kennedy had to make a decision. He had two options:

- take the risk of instant Soviet retaliation by initiating a military attack to destroy the missiles.
- blockade Cuba to stop more missiles from being transported (this will allow for negotiations between the two parties but may also incite a reaction from the Soviet Union).
- take no action, which would be interpreted as weak and push the Soviets to step up their aggression.

On the other side, Khrushchev had to choose between keeping his missiles in Cuba and bolstering the Soviet Union's standing in the Cold War or removing them and giving in to American pressure.

The game theory approach: anticipating the opponent's moves

Game theory became relevant in this situation because Kennedy and Khrushchev both had to plan a few steps ahead. They weren't just reacting to the current situation; they were also trying to predict how the other would respond to each possible action.

Kennedy couldn't afford not to do anything, even though he was aware that a full-scale military invasion of Cuba might start a conflict. He, therefore, decided on a compromise strategy: a naval blockade (sometimes known as a 'quarantine') of Cuba. This thought-out action meant to provide the Soviet Union with a means of escape without taking the situation too far. Kennedy was able to show his strength by demonstrating that the United States would not stand for Soviet missiles so near to its borders while allowing for some possibility for compromise.

Meanwhile, Khrushchev had to choose between lifting the siege and giving in. The idea of credible threats from game theory proved crucial in this situation. If Khrushchev thought Kennedy was serious about taking military action, then holding this position could result in nuclear warfare. But he could hold his stance if he believed Kennedy would finally yield.

The Nash equilibrium: finding stability

The Nash Equilibrium, as previously mentioned, is a situation in which both parties arrive at a result from which they can only both lose out. Both Kennedy and Khrushchev sought to achieve this balance during the Cuban Missile Crisis. They wanted a way to resolve the issue so that no one would feel pressured to take things farther or give in entirely.

The leaders were back and forth for several tense days before they eventually established a balance. In return for the United States agreeing not to invade Cuba, Khrushchev promised to remove the missiles from the island. Unbeknownst to the public at the time, the United States also gave consent in secret to remove its missiles from Turkey, which posed a threat to the Soviet Union.

The significance of game theory

The effectiveness of game theory in negotiations is aptly demonstrated by the Cuban Missile Crisis. Each party was functioning with partial knowledge; neither completely understood the other's goals or the possibility that they would carry out their threats. Game theory is based on the need to make judgments according to the assessment of what the other side would do.

What if Kennedy or Khrushchev had made a mistake in their calculations, believing the other would concede when they wouldn't? There might have been disastrous consequences. However, the fact that both parties adopted strategies that allowed for compromise while maintaining their strength allowed them to defuse the situation.

The same guidelines hold validity whether you're settling a personal dispute or engaging in commercial negotiations. Similar to Kennedy and Khrushchev, you need to look beyond the present moment and predict how the other person would act. Game theory is not limited to studying complex political situations; it also encompasses studying human behavior, strategy, and how to come up with solutions that benefit all parties. Therefore, keep in mind that planning ahead is essential the next time you find yourself in a high-stakes scenario. Furthermore, even while your conflict might not involve nuclear weapons, you can still win without resorting to violence by using thoughtful, strategic thinking.

CONFLICT RESOLUTION

Conflict is a part of life. Disagreements and tensions occur in all relationships – personal, professional, and political. The difficult part is figuring out how to settle these disputes amicably and constructively. Game theory demonstrates that managing and resolving conflicts in a way that is beneficial to all parties involved is just as important as competing and winning.

Thanks to game theory, we can think about conflicts in an organized way. A conflict can be seen as a game with players being the individuals or groups involved, strategies being the options available to them, and payoffs being the results each side will receive depending on their choices. By understanding these components, we can begin to sketch out possible solutions that are fair and, most importantly, sustainable.

Let's break it down. Imagine a conflict between two businesses for the same market. Each has an option: work together to achieve a win-win solution or carry on with their fierce competition, which can result in losses for both. Game theory helps both businesses see that while collaboration may result in better outcomes for both, continuous competition (or escalation) may ultimately hurt them.

People often concentrate on short-term advantages in conflicts, but game theory pushes us to consider the bigger picture. If one side wins today, what will happen? Does that breed animosity that becomes more problematic later on? Finding a win-win solution becomes, then, even more important than just winning.

Peace negotiations

Game theory is one of the best tools in peace talks, where lives are often at risk, and the stakes are great. It offers a framework for understanding why conflicts grow and how to de-escalate them, regardless of the nature of the conflict – a civil war or an ongoing international disagreement. Additionally, it provides resources for negotiating agreements that prevent either party from feeling defeated.

Think of a dispute between two nations. Every nation has a motivation to assert itself, display strength, and protect its interests. However, an escalation by either nation results in war, which worsens the position of both. According to game theory, peace is often the Nash equilibrium or the point at which neither nation gains from escalation. They can arrive at a solid solution where neither party wants to stray if they can both agree to give in just a little bit. Helping all parties see that peace is ultimately the best course of action, even though it may necessitate concessions.

The peace process in Northern Ireland serves as a practical illustration. Political and bureaucratic warfare produced an apparently unwinnable battle for many years. Each side made strong, opposing demands. However, after cautious discussions that might be understood as a game theory exercise, all parties concluded that continuing the violence was a lose-lose situation. A basis for sustainable peace was established in 1998 with the signing of the Good Friday Agreement, which turned peace negotiations into a means of mutual benefit.

The significance of communication and trust

Game theory also emphasizes how crucial communication and trust are to resolving disputes. Parties often remain silent because they are apprehensive about the other party's actions. Compromise is possible when there is clear communication of intentions and a mutually trusted relationship between the two sides.

Consider again the well-known scenario of the prisoner's dilemma, in which two suspects are detained and questioned separately. They will each receive a light punishment if they remain silent. When one betrays the other, the victim suffers while the betrayer escapes punishment. Should they both turn on one another, they will both face harsh consequences. The problem stems from a lack of trust because each prisoner must predict what the other will do. In real-world disputes, communication is key to ending the cycle of mistrust.

The lesson here is simple: when faced with a conflict, don't be scared to discuss your objectives with honesty and pay attention to what the other side has to say. Clear communication and trust pave the way for solutions that benefit all parties involved.

Finding a middle ground

The concept of striking a middle ground lies at the core of game theory as it relates to conflict resolution. It isn't a matter of one side triumphing over the other. Finding a solution where everyone feels like they've won something – even if it's not all they initially wanted – is the key. Because everyone is interested in seeing the outcome through to the end, this strategy results in more stable agreements.

Game theory provides guidance while negotiating corporate deals or resolving personal conflicts in your own life. It pushes you to consider the long-term effects of what you do, the motivations of the other party, and how you can both end up happy.

Although peace talks and conflict management are difficult processes, game theory offers a useful instrument for negotiating them. It serves as a helpful reminder to think strategically, to take other people's intentions and behaviors into account, and to work toward a solution that benefits all parties. Consider adopting these ideas to foster enduring peace and collaboration rather than concentrating only on winning or getting it your way. The final outcome? A dispute will be settled amicably through strategy and knowledge rather than force.

CHAPTER SIX

GAME THEORY IN INTERPERSONAL RELATIONSHIPS

A while back, a friend told me about her increasing stress at work. Her eyes showed the signs of exhaustion. She said, "I keep going faster and faster because I believe it will improve the situation. It's like I'm on a treadmill. However, things are just becoming worse. My mental health is failing, and my personal life is a complete wreck."

I stopped and gave myself some space to gather my thoughts. "Have you ever heard of game theory?" I enquired. Her eyebrow raised in curiosity. I explained: "Imagine if you approached your work and personal life as a holistic game rather than as separate entities. Gains in one area can occasionally translate into losses in another. And the overall outcome? It may be zero, negative, or even positive".

The game scenarios we saw in the previous chapters are often reflected in our personal and professional relationships. Consider this:

- It can seem like a zero-sum game in competitive settings, such as when competing with a coworker for a promotion;
- Collaborating with a friend to organize an event and sharing resources and thoughts results in a positive-sum game;
- Have you ever had a long-lasting disagreement where you both felt worse at the end? That is the way the negative-sum game works.

Relationships are more than simply one-on-one conversations; therefore, it's important to recognize these tendencies. Several games are going on at the same time, and you can get better results if you know which game you're playing.

RELATIONSHIP DYNAMICS

Communication and conflict resolution are essential in personal relationships. Game theory can provide new insights on how to resolve these conflicts – be they misunderstandings between partners, arguments with friends, or disputes within the family. Here are a few ways in which game theory might enhance your ability to communicate and settle disputes in interpersonal interactions.

Understanding the dynamics

The basic principle of game theory is that every encounter is a game in which each participant, or individual, has unique objectives, plans of action, and reactions. These games often entail a great deal of emotion and miscommunication in interpersonal relationships. By applying game theory to conflicts, you can improve your understanding of the dynamics at work and identify more efficient means of communication and problem-solving.

The basics of the game

To begin, let us consider the basics of game theory: players, strategies, and payoffs. The players in a personal connection are you and the other person. The payoffs are the results you individually receive from your decisions, and your strategies are the ways you decide to respond or behave in a given circumstance.

Consider a scenario where you and a friend disagree on what to do during the weekend. You may find yourself in a situation where neither of you is satisfied if you both dig in your heels and insist on sticking to your own plan. However, if you approach it from the standpoint of game theory, you might be able to reach a compromise that benefits both of you.

Communicating with empathy

Ineffective communication is one of the main obstacles to dispute resolution. Game theory emphasizes how crucial it is to understand the viewpoint of others. You may lose out on opportunities for compromise if your attention is solely on your own objectives and plans of action. Rather, try to view the circumstances from the perspective of the other person. What are their objectives and worries? How can you modify your approach to deal with these?

For instance, it's critical to understand your partner's financial concerns and objectives if you are having financial arguments. You can make a plan that meets your requirements and theirs by accepting their worries and implementing them into your own plan.

Finding common ground

Finding a Nash equilibrium, in which neither party gains anything by changing their stance if the other player's strategy stays the same, is a desired result in game theory. This idea extends to establishing a middle ground in interpersonal relationships where both sides feel respected and heard.

Imagine that while you and your family are organizing a trip, everyone has different ideas for where to go. Rather than having everyone adhere strictly to their preferences, apply game theory to identify a solution that combines aspects of each person's suggestions. Perhaps you can divide the trip into segments to accommodate the various interests, or perhaps you can go to a place where there are activities that everyone enjoys.

Using the prisoner's dilemma

In the classic game theory situation known as the Prisoner's Dilemma, two people must decide whether to work together or turn on one another. They both gain if they work together. The one who betrays gains more when the other one cooperates. They both lose out if they both turn on each other.

This dilemma often shows itself in trust issues in relationships. When you and a friend have a falling out and neither of you chooses to make amends, the friendship suffers. However, you will both be in a better place if you decide to talk to one other and extend forgiveness. By using game theory in this context, we can see that collaboration and trust often produce greater results than hostility and revenge.

Strategic compromise

Making calculated concessions rather than focusing on winning or losing might sometimes be the key to resolving a dispute. This is known as striking a balance in game theory, where each party receives some of what they desire.

For example, if you and your partner are at odds over who should do what around the house, try to work out a timetable that works for both of you. Even though you might not achieve everything you desire, you both gain from a more peaceful living environment when you compromise.

The importance of trust

Trust is essential in both game theory and interpersonal relationships. It's easier to come to agreements and settle disputes when you have trust in the other person's motives and behavior. Being dependable, honest, and transparent in communication are all necessary aspects to establish trust.

When handling a dispute, concentrate on establishing or restoring confidence. Communicate your feelings honestly, and give the other person your whole attention. This strategy can lay the groundwork for improved dialogue and more successful dispute resolution.

TRUST AND COOPERATION

Any healthy connection, whether it be with a partner, friend, family member, or coworker, is built on trust and cooperation. It's challenging to create meaningful connections without trust, and resolving disputes becomes more difficult in the absence of cooperation. Game theory provides some helpful insights into how relationships function in terms of trust and cooperation, as well as how to build these qualities over time.

The role of trust in relationships

In game theory, decisions that result in cooperation or conflict often revolve around trust. Again, with the prisoner's dilemma, two individuals have to choose whether to work together or turn against one another in this situation. They both gain if they decide to work together. However, in the event that one betrays and the other assists, the betrayer benefits more, and the cooperator suffers. If one turns on the other, they both lose.

Let's apply this principle to relationships now. Imagine that you and your partner are debating how to resolve a conflicting matter. You'll likely find a solution that benefits you both if you have mutual trust and cooperate. However, the relationship may suffer if one of you starts to distrust the other's motives and decides to act (or betray the other) in their own self-interest. Furthermore, neither of you gains if you decide to part ways.

Gradually, trust is established through minor, regular behaviors. According to game theory, every constructive interaction – such as keeping your word or being there for someone when they need you – builds trust. However, actions that undermine trust, such as betrayal, breaking your word, or dishonesty make cooperation more difficult.

How to build trust using game theory

According to game theory, repeating games helps to foster trust. This implies that rather than a single encounter, you will have numerous opportunities to make decisions, converse, and establish trust. Every time you choose to be cooperative and honest with the other person, your mutual trust is reinforced.

For instance, suppose you are resolving a conflict with a friend. You're building trust when you decide to listen intently and look for points of agreement rather than focusing solely on trying to 'win' in the argument. Your friend is more inclined to work with you in the future since they will remember these encounters over time.

In relationships, being consistent fosters trust. It's similar to repeatedly playing a game where both parties discover that cooperation produces greater results than confrontation. It's not necessary to always win or get your way in arguments; what is important is demonstrating your reliability to the other person.

The value of cooperation

Another essential component of relationships and game theory is cooperation. When two individuals work together, they both get closer to realizing their goals. Consider it a win-win situation. Cooperation in interpersonal interactions refers to teamwork, listening to one another's needs, and coming up with solutions that satisfy both sides.

According to game theory, tit-for-tat strategy is the most effective means of guaranteeing long-term cooperation. This essentially indicates that the likelihood of reciprocation is high when one party cooperates with the other. Should one turn disloyal, the other may do the same. This dynamic is common in relationships: if you make an effort to get to know and support the other person, they'll probably do the same for you. When one partner consistently selects self-serving actions, the relationship may begin to deteriorate.

Let's say that one partner in your relationship does the majority of the work. Resentment grows over time, and the other party may become less willing to collaborate. On the other hand, cooperation strengthens the bond and facilitates future collaboration when both parties actively cooperate and share duties.

Maintaining trust and cooperation over time

Relationships, much like repeated games in game theory, require consistent effort to reinforce cooperation and build trust. Trust is built gradually through reliable and positive interactions—it's not a one-time achievement. Each interaction offers an opportunity to choose collaboration or self-interest.

For example, imagine you and your partner are discussing financial decisions. To foster a productive outcome, try focusing on practical actions: be transparent about expenses, actively listen to each other's concerns, and work together to establish shared financial goals. Trust can quickly erode if one person dominates the conversation or keeps details hidden, so fostering an open and balanced dialogue is key.

This approach applies beyond romantic relationships; in friendships and professional partnerships alike, maintaining cooperation means consistently sharing responsibilities, communicating openly, and respecting each other's needs. In this way, you establish a foundation of trust that can sustain even during challenging times.

Repairing trust when it's broken

In game theory, miscommunication or poor decisions can occasionally lead to the betrayal of trust. This also holds true for real-life relationships. Mistakes happen; perhaps someone disappointed you or abandoned you when you needed them. Even while it can be challenging to regain trust after it has been damaged, game theory shows that it's not impossible.

Over time, consistent and cooperative actions are the most effective method to rebuild trust. Consider a player who makes a poor move in a game but always opts for collaboration in subsequent rounds. The other player eventually comes to trust them once more. This entails admitting where things went wrong in a relationship, offering a real apology, and then proving to the other person that you're determined to rebuild trust through your actions.

Long-lasting, meaningful relationships require trust and collaboration, and game theory provides a basis for understanding how to establish and preserve these components. Each conversation you have with others can either build or diminish their trust. Strong, enduring, and mutually beneficial relationships can be built by continuously choosing cooperation and being dependable.

Thus, consider your relationships as a sequence of decisions that result in many outcomes rather than as a manipulative game. Long-term relationships will benefit from your increased choice of cooperation and trust.

SOLVING COMMON RELATIONSHIP DILEMMAS WITH GAME THEORY

Let's examine how game theory can help in resolving a few common relationship problems. Game theory can offer you new insight into how to handle difficult situations, whether they involve a partner, friend, or family member. You'll begin to notice that a lot of the problems we encounter in relationships have patterns that, given the right approach, may be recognized and resolved.

1. The dilemma of who apologizes first.

Suppose you just got into a fight with your partner. You both feel hurt, and there's this awkward standoff – who will apologize first? While neither of you wants to appear weak by apologizing, you also don't want the dispute to continue. In the case of the classic game, two players are heading in the same direction. Whoever steers from their path first and avoids the collision appears to be giving up. In a relationship, neither party wants to be perceived as the one to 'give in,' yet they both want the conflict to stop.

Game theory solution: in this case, the game theory approach would be to find a mutual advantage rather than to concentrate on winning. Offering an apology first is really a helpful gesture. You can let your partner know that you cherish your relationship more than your disagreement by apologizing. You steer to avoid an accident, but in the process, you both get peace and resolve, which is better. You might discover that using this strategy gradually builds trust. Your partner discovers that since you're both aiming to keep your relationship solid and well-maintained, they don't need to back down or worry about 'losing' the argument.

2. Deciding on plans together

Have you ever faced the dilemma of "what should we do this weekend?" with a partner or friend? Maybe one of you wants to go out for dinner, but the other would rather spend a quiet evening at home. Everyone has their own preferences, and if you both stubbornly stick to your decisions, nobody will be able to get what they desire. This is comparable to the coordination game, in which the participants' preferences differ, yet they still gain from coordinating their actions. Here, figuring out how to align your interests is crucial.

Game theory solution: to use a game theory analogy, compromising is the optimal course of action here. Perhaps you decide to go out to dinner this weekend and have a peaceful night at home the following weekend. Over time, you can make sure that both parties are happy by switching between preferences. In this manner, you collaborate to ensure that you are both happy in the long run rather than fighting to win every time. Additionally, this strategy builds trust. When your partner sees that you're willing to make concessions, they'll be more inclined to follow suit in the future, which will strengthen and improve your relationship.

3. The silent treatment standoff

Let's discuss the infamous silent treatment. Sometimes, following an argument, whether it's between friends or in a romantic relationship, both parties stop talking. It gets difficult to break the silence the longer it lasts as each person waits for the other to initiate contact. This is a variation of the Prisoner's Dilemma. The relationship suffers when you both remain silent. If both people don't speak, the harm is multiplied. If one person breaks the silence, they can worry about being 'punished' for being weak.

Game theory solution: realizing that ending the silence is an act of cooperation is the first step in finding a solution to this problem. Reaching out first shows that you are prepared to move forward, even if you believe the other person is more at fault. This can be understood as a tit-for-tat strategy in game theory, in which one player breaks the silence with a cooperative intention, and the other player responds by joining in the conversation. Instead of allowing the lack of communication to become a bigger issue, you're fostering cooperation by taking the initiative. It demonstrates to the opposing party your commitment to finding a solution rather than escalating the dispute.

4. Balancing chores and responsibilities

There is often a dilemma in all relationships – romantic or familial – about who handles more household chores. It's possible that you feel like you're doing most of the work, and the other person feels like they're chipping in equally. If this is not handled appropriately, it might cause resentment and arguing. This situation can be analyzed through the lens of game theory's bargaining games. Based on their perceptions of each other's actions, both parties are bargaining on how much work they should each put in.

Game theory solution: the best course of action in this situation is to speak and bargain honestly. Sit down with the other person and discuss with honesty the distribution of chores rather than allowing resentment to build. This is similar to reaching a Nash equilibrium in game theory, which is the point at which neither of you has any reason to change your behavior because you have both reached a fair balance. You are working together to create a solution where both parties are happy with the arrangement by talking about expectations and deciding on a division that feels equal. This keeps things from getting out of hand and strengthens the bond between people.

5. Handling financial decisions together

In any relationship, it might be challenging to discuss money. One individual may prefer to save money while the other is more at ease making purchases. This can result in disagreements on how much money to set aside for holidays or other major expenditures or how much to save. This is comparable to a mixed strategy game in which the two players must compromise to prevent conflict despite having differing preferences.

Game theory solution: game theory indicates that by combining their strategies, both parties can find a solution in this situation. You may wish to set aside a certain amount of your income for savings and another for personal expenses. Instead of making one individual give up their plan entirely, you're able to create a hybrid solution that meets the needs of both by combining their preferences. In the long term, this form of compromise strengthens financial stability and lowers stress levels since both parties feel their demands are being taken into account.

PART 3

ADVANCED APPLICATIONS AND TECHNIQUES

CHAPTER SEVEN
EMOTIONAL GAME THEORY

As the Disney-Pixar animated film *Inside Out* (2015) masterfully illustrates, emotions drive most, if not all, of our actions. However, economists paid little attention to emotions and how they influence human behavior for a very long period. Even in situations where game theory appears to be reasonable, emotions have a substantial influence on decision-making. A player's decisions and, ultimately, the results of the game may be influenced by feelings such as fear, greed, anger, or other negative emotions.

In the context of negotiations, for example, one party may have a strong sense of pride and be unwilling to give in, even if doing so would result in a win-win situation. Conversely, even in situations where collaboration isn't the best course of action from a strictly strategic one, having empathy or sympathy for the other person may inspire cooperation.

By understanding how emotions function in game theory, we can create more effective incentives and strategies that take these emotional aspects into consideration. This can lead to better results and more efficient decision-making.

BEHAVIORAL GAME THEORY

A common belief among people is that emotions should be avoided since they impede decision-making. Instead of feeling them, processing them, and understanding their meaning, they might ignore or repress them. When it comes to making decisions, people would rather use reason than emotion.

Emotions, however, are valuable. It seems that we would be inactive and achieve nothing if emotions weren't there to inspire and push us. Since emotions are meant to elicit certain responses, our emotional states greatly influence the decisions we make. Emotions condense experiences fast and assess them to guide our choices, enabling us to act swiftly during situations of emergency.

Although emotions guide us, they are primarily motivated by our innate need for survival. As a result, emotions typically convey their signals below our conscious level. It's vital to remember that emotions are not very precise due to their speed and their survival instinct. Their effectiveness and quickness make up for their lack of precision and complexities. Because of this, the emotional system often raises false alarms, forcing us to reconsider our course of action and determine whether it is suitable for the given circumstance.

Recent studies have shown the importance of emotion in logical decision-making. When there are no emotional cues, decision-making is nearly impossible. Our well-being may depend on our ability to understand and interpret our emotions while combining them with reason to make the right decision. Emotional cues should be taken into account and processed, but we also need to assess whether our reactions are appropriate for the given circumstance.

How do you use emotions to make good decision-making? The following actions can help you use emotions to your advantage while making decisions:

1. **Embrace your emotions**

Don't suppress or disregard your emotions. First, try to recognize and understand them. Give yourself a chance to acknowledge your emotions and the reasons behind them. Healthy decision-making requires this deliberate self-examination process since emotions can alter our opinions and judgments.

2. **Remember the emotional bias**

Due to their survival instinct, emotions tend to form preconceptions that affect how we see the world and how we interpret what happens. Remember that safety is more important to the emotional brain than accuracy. As you debate its message, pay attention to its alarm signal.

3. **Regulate your emotions**

Emotions affect our ability to reason, especially when they are strong. Our ability to think objectively and logically can be hindered by intense emotions, which can also impair judgment. Because of this, it's critical that we regulate our emotions so that they are appropriate and balanced for the circumstances.

4. **Use your emotions as guidance**

When it comes to directing you toward those pursuits that are most meaningful to you and/or aligned with your values, your emotions can act as a compass. But it's crucial to resist allowing emotions to control completely your judgment. Be mindful to strike a balance between rational thinking and emotive insights.

5. **Hold onto your logical thinking**

Enlisting the assistance of the rational intellect is crucial. By doing this, you shift from a fast-moving, instinctive, and unconscious system to a slower-moving, more regulated, logical, and conscious system. You transition from an emotional system that is reactive and impulsive to one that is strategic, adaptable, and introspective.

6. **Think about the context**

Examine the current circumstance and keep in mind that the context may have an impact on emotions. Your judgment may be affected by emotions resulting from personal prejudices or past experiences. Keep the past and the present separated and concentrate on the important elements.

7. Collect relevant information

Though they should be supplemented with factual knowledge, emotions can offer insightful information. Before making significant decisions, take your time to acquire vital facts. Weigh the pros and cons of your selections to make the best choices possible.

8. Be mindful

The secret to mental harmony is mindfulness. The uncontrolled mind can become delusional, giving in to irrational desires, passions, and emotions. We can employ reason to interpret our emotions by practicing mindfulness, which helps us become conscious of them. Use your emotions as a doorway to access a higher state of awareness.

9. Develop compassion

Developing compassion in decision-making is a powerful means of producing more moral, ethical, and well-rounded decisions that take everyone's welfare into account. Compassion helps us choose acts that will benefit both ourselves and others.

10. Develop emotional intelligence

It is said that emotional intelligence is the ability to recognize and regulate your emotions. Emotional intelligence is mostly composed of motivation, self-regulation, empathy, self-awareness, and social skills. Gaining emotional intelligence skills will enable you to make decisions based on your emotions without letting them rule you.

11. Reframe the situation

Reframing is the deliberate alteration of one's interpretation of an emotionally charged event in order to lessen unpleasant emotions. By intentionally thinking loving thoughts and showing compassion to both yourself and other people, you can change the way you interpret an experience.

12. Extend your viewpoint

You are not sidetracked by minor problems or whims when you have a clear vision and are committed to your ultimate goal. Emotions can be directed toward harmony and peace by identifying and pursuing your deepest long-term goals. It will enable you to see that, regardless of the result, the choice that is motivated by your values is the right one.

TRUST, RISK, AND COOPERATION

Our emotional states have a significant impact on how we handle risk-taking and trust in both personal and professional contexts. It's easy to concentrate on the logical aspects of game theory – the strategies, results, and payoffs – when discussing it. However, in actuality, emotions have a significant impact on our decision-making process, often tilting the scales in ways that may not initially appear reasonable.

Trust and emotions in personal relationships

Consider the concept of trust in your own life. Whether it's with a friend, lover, or family member, you constantly have to make riskier judgments while you're in a relationship. For instance, you might depend on a friend to fulfill a commitment or trust your partner with a private secret. These are tiny but powerful moments that demand you to take a risk based on your emotions and past interactions with that individual.

According to game theory, trust often involves accepting a calculated risk. You balance the possibility of being let down with the possible reward of trusting someone, such as strengthening your relationship. An excellent example of this is, once again, the prisoner's dilemma, in which two individuals must decide whether to trust each other or not.

Emotions muddle this calculation in real life. Even in situations where cooperation could result in a better end, you can be reluctant to trust someone if you have previously been harmed or deceived. Your decision to take a risk or not may be influenced by your feelings of vulnerability and past experiences, which may make you more cautious. Positive feelings, however, such as gratitude, love, or affection, may increase your likelihood of trusting even when the risk is great.

Risk-taking in professional contexts

Risk-taking is a common occurrence in the business sector, whether you're settling on a partnership with a competitor or negotiating a pay raise. When your job growth or livelihood is at stake, the threats can seem considerably more real.

This is where game theory's consideration of emotions comes in. Consider yourself and a supplier in a negotiation game. You run the risk of destroying the relationship or losing the contract entirely if you take a firm stand and demand reduced prices. However, you may lose money if you're overly lenient. Although a thorough analysis of the costs, advantages, and probabilities would be the sensible course of action, your emotions during the negotiation also play a role in your decision-making.

When you're feeling powerful and in charge, you can be more willing to take chances and negotiate better terms. You could be cautious if you're uneasy or insecure. Our willingness to take chances is often determined by emotions such as fear, confidence, enthusiasm, or dissatisfaction; these feelings can sometimes result in better outcomes and, other times, in missed opportunities.

The game theory angle

From the standpoint of game theory, it is essential to understand how emotions affect risk and trust. It needs more than just reasoning or the figures to make decisions, whether in your personal or professional life. Your approach and how you perceive risk can both be affected by emotions.

For a moment, let's return to the prisoner's dilemma. When negotiating a partnership in a professional situation, if your gut tells you to distrust the other person (maybe as a result of a negative experience in the past), you may decide to betray or leave the conversation even though working together could produce a better outcome overall. On the other hand, even when there is a small amount of risk involved, you might be more likely to trust and collaborate if you are upbeat or have a close relationship with the other person.

The trick is to strike a balance between your gut feeling and the rational explanation that game theory offers. Although emotions play a role in decision-making, you can't dismiss them; instead, you can make better decisions by understanding their influence.

Striking a balance

Emotions often serve as a filter in both personal and professional contexts when evaluating risk and trust. Realizing that emotions are a genuine and powerful factor in decision-making, even as logic and strategy play a major part, is a crucial component of game theory.

What's the most important lesson you can learn from this? When faced with decisions that need trust or carry risk, stand back and consider how your emotions are influencing your selections in addition to the obvious results.

Making smarter, more balanced judgments can result from integrating strategic game theory ideas with emotional awareness, whether you're negotiating a difficult deal at work or handling a problem in your personal life.

You may use game theory as a useful tool for navigating real-life circumstances where emotions and risk are always at play, rather than merely as a set of abstract concepts, by detecting and accounting for your emotional state. This will help you improve both your personal relationships and professional outcomes.

CASE STUDY

Let's have a peek inside the world of professional athletics, where pressure, stakes, and personal emotions continuously influence decision-making, and emotions are often strong. LeBron James' 2010 trade from the Cleveland Cavaliers to the Miami Heat is a prime illustration of how emotions play a crucial role in decision-making. This was a highly emotional issue involving fans, loyalty, legacy, and individual aspirations; it was more than just a sports decision.

What happened

The Cleveland Cavaliers were home to LeBron James for the first seven years of his NBA career, making him maybe the best basketball player of his generation. The fans expected him to lead the club to a title since he was a hometown hero. However, he was forced to decide whether to remain faithful to Cleveland or take his skills elsewhere in the hopes of winning a championship after failing to win one on multiple occasions.

LeBron was up against a traditional decision game from the standpoint of game theory. He had a few options: join forces with other elite players and try to win more titles with the Cavaliers, or go and hope they could assemble a stronger team elsewhere, such as the Miami Heat. The possible results were obvious: if he stuck around and took home a championship, his reputation as a devoted fighter would be solidified. However, if he persisted and lost, he would run the risk of being remembered as a fantastic player who was unable to win the big one. Taking charge of his future by relocating to Miami came at the expense of his status as Cleveland's favorite hero.

The emotional layer

This decision's emotional significance is what made it so intriguing. For LeBron, it was about devotion, his own identity, and the expectations of millions of fans, not just the hard, cold logic of winning championships. The choice to leave Cleveland was emotionally charged on both sides.

His hometown was devastated, and supporters felt betrayed. The decision was treated like a circus by the media, which examined every scenario. Making the choice in front of the entire world on live television (during the infamous show named *The Decision*) only intensified the emotions, as LeBron himself later acknowledged.

LeBron's decision to join the Miami Heat was a pivotal moment in American basketball, particularly in the NBA. Although he felt pressure to cement his legacy as a champion and had personal aspirations, LeBron also felt a strong connection to his hometown team, the Cleveland Cavaliers. From a game theory perspective, joining the Heat was a strategic choice, as it maximized his chances to win championships by teaming up with other star players in a highly competitive league. With Miami, he went on to win two NBA championships, solidifying his status as one of the all-time greats in basketball history. This decision highlights how, in professional sports, long-term goals and legacy can sometimes outweigh loyalty to a single team.

How emotions shaped the outcome

There was a tremendous outcry against his decision. LeBron received harsh criticism, with many fans considering the action to be a betrayal. It was impossible to calculate the supporters' emotional response using only rationality. It's the human aspect of the game, and it's when emotions begin to have an erratic effect on results. In this instance, there was an immediate emotional cost to his reputation, but in time, people understood his decision, particularly after he returned to Cleveland and helped the team win a championship.

LeBron might have handled the matter differently if he had completely disregarded his emotional ties to Cleveland. He could have decided not to do the live broadcast, or he could have explained his decision in a way that honored his relationship with the fans. Though it resulted in a difficult start, LeBron was able to successfully negotiate the balance between his personal allegiance and his career aspirations by taking his emotions into account.

The game theory perspective

LeBron's decision was a typical payoff matrix scenario from the perspective of game theory: stay and run the risk of long-term unhappiness, or leave and risk immediate backlash in exchange for the possibility of a bigger return. There were more layers of complexity due to the emotions involved, both his own and the supporters'. According to game theory, choices aren't made in a vacuum. Even the most rational minds are influenced by emotions such as loyalty, remorse, anxiety, and peer pressure.

You may not be choosing which NBA team to join in your personal life, but you undoubtedly have to make difficult choices with a lot on the line. Emotions are always a factor in life decisions, whether they involve quitting a job, breaking up with someone, or changing drastically. Game theory can help you analyze the rational side of your decisions, but it's crucial to keep in mind that emotions – your own and those of those around you – will affect how things turn out in the end.

LeBron's choice shows us that sometimes there are emotional consequences to even the most logical option. But just like LeBron did when he decided to pursue his dreams on his terms, you can prepare for it, manage it, and ultimately make the best decision for yourself if you are aware of the emotional environment.

CHAPTER EIGHT
GAME THEORY IN EMERGING FIELDS

Game theory is an intriguing area of study with practical applications in a number of industries, including social media, e-commerce, and technology. It offers a framework for understanding the actions of logical agents in strategic scenarios where one's decision will have an impact on the decisions made by others. In this chapter, we shall examine the real-world uses of game theory in emerging fields. We can improve the outcomes of our interactions with people and make better strategic decisions by knowing how game theory can be applied in various contexts.

TECHNOLOGY AND INNOVATION

Game theory is more important than ever in today's world, particularly in areas like artificial intelligence (AI), blockchain, and cybersecurity. Though the mathematical concepts of game theory may not immediately spring to mind when you think about algorithms or digital transactions, they are fundamental to how networks work, how decisions are made, and even how confidence is established in decentralized systems.

Artificial Intelligence (AI) and game theory

Artificial intelligence systems often have to make decisions when engaging with people or other systems. Consider autonomous vehicles, for example. In addition to navigating roadways, they must communicate with other autonomous cars and human drivers. Game theory is relevant in this situation because these systems must predict other people's actions, such as a car abruptly changing lanes or a person quickly crossing the road.

AI can make better decisions by using game theory, which takes into account other people's expected decisions. It helps AI developers create systems that 'think' about the effects of various actions and simulate various scenarios. Game theory, for instance, can be used to optimize traffic flow in AI-powered traffic management systems by predicting how other cars or drivers will respond to particular signals or changes in the road. This is critical because artificial intelligence (AI) goes beyond real-time response to include result prediction and optimal decision-making in multi-agent settings.

Blockchain and decentralized networks

A decentralized trust model is fundamental to the success of blockchain technology. Traditional systems rely on centralized trusts, such as payment processors or banks, to verify the legitimacy of transactions. All players in a blockchain share a certain amount of trust, and game theory is crucial to maintaining the proper and equitable operation of the network.

Take Bitcoin mining as an example. Here, miners (players) compete to solve cryptographic riddles to validate transactions and safeguard the network. Despite the unpredictability of the payouts, game theory explains why these miners devote their time and computational resources to the task. Each miner acts based on a payoff structure: the possibility of receiving a reward (Bitcoin) against the costs (electricity, hardware). A person who attempts to cheat or jeopardize the network will either receive nothing at all or, worse, ruin their own interests due to the meticulously crafted incentive structure.

Blockchain systems could be easily vulnerable to malevolent behaviors, low participation, or instability without these game-theory-driven incentives. Each player in the network knows the rules of the 'game' and plays by them to optimize their own results, which eventually contributes to the proper operation of the entire system.

Cybersecurity and game theory

Game theory is particularly effective in the subject of cybersecurity. Imagine that a company is trying to protect its networks from possible attacks from hackers. The hacker and the organization are engaged in a strategic competition in which they are trying to outwit one another. The company needs to learn how to safeguard its resources because the hacker is always looking for openings.

Game theory models adversarial behavior, which aids cybersecurity experts in anticipating and thwarting assaults. We call this adversarial modeling. It all comes down to anticipating potential attack tactics and building defenses in response. Game theory, for example, is used to determine how resources should be distributed to safeguard the most important or fragile components of a system, considering the possibility of assaults and the possible harm they could do.

Game theory can also be used to predict user behavior in phishing attacks, which are attempts by hackers to fool users into disclosing sensitive information. It enables security specialists to create countermeasures that are more effective by studying how to train users to avoid being misled and modeling various attack situations.

Knowing the functioning of game theory aids in understanding how decisions are made in these complex systems, regardless of whether you work in technology or business or are just a consumer of these tools. AI aims to make machines more intelligent. Building confidence in blockchain eliminates the need for middlemen. Additionally, it's helping businesses keep one step ahead of cybercriminals. Recognizing the importance of game theory will help you better explore and understand the technologies that are defining our future. Game theory is the invisible hand that guides the evolution of these tech-driven areas and understanding the 'rules of the game' offers you a significant advantage whether you're developing systems, choosing a course of action, or just safeguarding your data.

MODERN APPLICATIONS

Game theory is not limited to economists and mathematicians; it's also rooted in the daily decisions that companies make in the areas of social media, e-commerce, and product creation. When it comes to pricing strategies, when to introduce new products, and how to interact with customers on social media, game theory helps firms anticipate competitive behaviors and customer feedback, which eventually results in more informed decisions.

Let's examine these crucial areas more closely.

E-Commerce: pricing and competition

When it comes to e-commerce, price is crucial. Your product's pricing may make it or break it; therefore, it's never enough to base it only on what you believe is right. You must take into account the demands of your customers, your competitors, and the general dynamics of the market. Game theory can help in this situation.

Consider internet retailers such as Amazon.com. When numerous sellers provide identical products, they are participating in a pricing game. Each seller must choose their price, aware that if they charge too much, buyers may choose a less expensive competitor. A pricing that's too low could result in their not making enough money or even in a losing pricing war for all parties.

Game theory helps sellers in determining a pricing equilibrium in this situation. They examine pricing strategies used by competitors, consumer purchasing habits, and industry developments. For instance, a seller may choose to launch a temporary promotion in order to draw in more business without going all out in a price war after realizing that their competitor's prices have been the same for too long. The aim is to make calculated decisions that maintain competitiveness while balancing profitability.

Social media strategies: engagement and content

When you post on social media, especially as a brand or business, you're not just posting in a vacuum. You are in competition for the interest, loyalty, and attention of your target audience. Brands can use game theory to help them strategically consider how to position themselves in the crowded social media space.

Imagine that your company is starting a fresh Instagram marketing campaign. You are aware that your competitors will probably be promoting their own content concurrently. How do you differentiate yourself? Game theory pushes you to concentrate on your own advantages while keeping an eye on your competitors' moves. For instance, you may try posting over the weekends when fewer brands are active, especially if your competitors are usually active on weekdays. Alternatively, you may use movies or interactive content to set yourself apart if they're only using static images. The goal is to maximize your own influence while intelligently reacting to others' actions. It's not enough to only imitate your competitors; you also need to use your understanding of the game to develop a beneficial plan.

Additionally, brands employ game theory to predict followers' responses to particular kinds of content. You can create anticipation and engagement by posting at the same time every day or by utilizing a specific kind of content (such as tutorials or memes), which will eventually increase likes, shares, and visibility.

Product development: anticipating market moves

When businesses develop new products, they aren't doing it in isolation. The demands and preferences of customers are evolving, and competitors are also making an effort to stay one step ahead. Understanding game theory is essential to making wise judgments about which products to create and when to launch them.

Picture that, as a company, you are working on a new smartphone. You are aware that your competitors are working on their own new products, and releasing yours too soon or too late could reduce sales. To determine the best launch window, you examine competition schedules, market trends, and customer input using the matrix of game theory. Launching your product too soon could leave it without important functionality. However, if you hold for too long, a competitor may end up controlling the market.

Classic examples of businesses utilizing game theory in product development include Apple and Samsung. Each company closely monitors the features and release schedules of the other. In order to get an advantage in a different market, Samsung may hasten the release of their Galaxy phone with a focus on camera quality if Apple is planning to release a new model with a longer battery life. Beyond just scheduling, game theory helps companies prioritize features according to customers' desires and potential competition offerings. You may decide to spend on premium features in order to draw in a new market segment if you are aware that your competitor is prioritizing price.

Game theory gives you an advantage whether you're an independent company owner, a prospective business owner, or just interested in how decisions are made. It teaches you how to think strategically about what other people are doing (competitor businesses or customers) as well as your own actions. Understanding the rules of the game will help you make better judgments in product development, social networking, and e-commerce.

The next time you shop online, browse Instagram, or check out the newest gadget release, keep in mind that a lot of game theory is going on behind the scenes, and those calculated decisions are influencing the environment in which you live.

ETHICAL CONSIDERATIONS

Even while game theory is a strong tool for developing strategies in fields such as product development, social media, and e-commerce, it's crucial to consider the ethical considerations of its application. Ultimately, the goal of game theory is to obtain the upper hand by strategically planning plays based on potential outcomes from opponents. However, as these strategies start to affect decisions, emotions, and day-to-day activities, we need to consider whether this is always the best course of action.

Manipulation vs. strategy

The difference between deception and clever planning can occasionally become hazy in business. Game theory teaches us to anticipate other people's actions and respond appropriately, but this might result in tactics that aren't always ethical or just. Is it fair, for example, for a business to use game theory to start a pricing war with the sole goal of driving out smaller competitors? When game theory is applied too aggressively, it can often be detrimental to competitors as well as consumers who depend on a fair market for affordable prices.

This also applies to social media strategies. Brands often employ game theory in order to boost engagement. They upload content at the appropriate times and craft messaging that they know will foster interactions. However, what would happen if those strategies began to control people's emotions or attention spans? Our algorithm-driven society can occasionally force companies to put engagement ahead of well-being, raising ethical concerns about the usage of such strategies if they encourage addictive habits or distort the truth for their own benefit.

Privacy concerns

Privacy is another ethical factor to take into account, especially in tech-driven and e-commerce areas. Since data is king, game theory often uses vast volumes of data to predict action. Businesses monitor user behavior, past purchases, and even browsing patterns to determine the best course of action. However, as customers, do we always know how much of our private data is being used to create these strategies? There is a fine line between violating someone's right to privacy and using data to improve marketing experiences.

Consider the scenario where you are online shopping, and all of a sudden, advertisements for the product you were just looking at appear on every website you visit. It's essentially game theory in action, using your past purchases to predict what you could buy next, although occasionally it feels intrusive. Companies must guarantee that they are protecting consumer privacy and being open and honest about how they employ consumer data to make strategic decisions.

Creating a fair playing field

Game theory also forces companies to continuously look for ways to gain a competitive edge. This encourages creativity, but it can also lead to imbalances. Smaller brands may find it hard to compete when larger, more resourceful enterprises employ game theory to perfect their pricing or marketing techniques. This may lead to monopolistic conditions when a small number of firms control the market, restricting customer choice and stifling the creativity of smaller entrepreneurs.

For instance, larger businesses may have access to superior data in tech-driven industries like blockchain and artificial intelligence, which enables them to create products that outperform those of their smaller competitors and make better decisions. Although competition is inevitable in business, it's crucial to think about whether these strategies actually help to diversify and innovate the market over the long run or if they just help to consolidate the power of a small number of people.

The human factor

How much does game theory reduce human interactions to cold, calculating maneuvers? This is perhaps one of the largest ethical considerations when implementing it, especially in quickly expanding fields like social media, product development, or even personal relationships. While human behavior in real life is significantly more complex than game theory, game theory believes that players operate rationally, constantly aiming to maximize their benefit. Decisions made solely from this perspective may overlook the significance of emotions, morals, and values.

For instance, applying game theory to predict someone's next move in a relationship – personal or professional – can create a strategic 'game' out of every meeting rather than a sincere bond. Imagine using game theory to determine the best time to schedule a challenging conversation with your partner based only on what will benefit you the most. While it could work in the short run, is it fostering the development of an honest and respectful relationship?

A balanced approach

The most important lesson is that even while game theory provides fantastic resources for strategic thinking, it should always be used ethically and consciously. It goes beyond simply winning over or outwitting the opposition. Businesses, leaders, and individuals must exercise mindfulness of the impact of their strategy on others, including consumers, competitors, and the wider market.

When applying game theory properly, one must put justice, openness, and other people's interests first. Whether you're crafting the next big social media campaign or negotiating a supplier agreement, asking yourself "Is this the right thing to do?" should be as important as asking "What's the smartest move?" After all, the most successful strategies are the ones that win not just today but in the long run by building trust and maintaining a fair, competitive environment.

CHAPTER NINE
GAME THEORY'S CROSS-DISCIPLINARY IMPACT

Game theory has several uses in various academic fields. It offers important insights into the dynamics of conflicts, negotiating techniques, and decision-making processes by examining the strategic interactions between individuals or groups. In this chapter, we shall examine how game theory concepts can be used in these fields to understand and simulate human behavior.

GAME THEORY AND EVOLUTIONARY BIOLOGY

Aside from being a useful tool in commerce and negotiations, game theory is also essential to understanding nature and the evolution of life. The ways in which plants, animals, and even bacteria behave are often similar to how game theory analyzes strategic decisions. This is because existence in the natural world resembles a high-stakes game in which participants must continually cooperate or compete in order to survive and procreate.

Survival as a strategy

In order to survive, animals in the wild must choose when to cooperate, when to compete, and how to navigate their surroundings. Our instincts, which evolved over millions of years, often arrive at these decisions unconsciously. However, upon closer examination, these actions conform to the same reasoning that game theory uses to evaluate human decisions.

For instance, let's discuss cooperation. Picture a community of animals surviving in a hostile environment with limited access to food and water. Like food sharing or group hunting, cooperation turns into a useful strategy. It isn't because the animals bargained like people do; rather, it's because, in the long run, cooperative animals have a higher chance of surviving. This is referred to as a win-win strategy in game theory when cooperating helps all parties involved.

The prisoner's dilemma in nature

It is true that one of the most famous game theory scenarios, the prisoner's dilemma, exists in nature? The basic concept is the same: both players have the option to work together or turn against one another, and their decisions will determine the result. This dilemma arises in nature when animals have to choose between acting selfishly or sharing resources.

Take vampire bats as an example. These bats often exchange food by spitting blood to fill the stomachs of other hungry group members. However, there's always a chance that some bats won't give back when it's their turn. When a bat declines to contribute, it is essentially betraying the community by abusing the cooperative environment without making any contributions. However, consistent cooperative bats have a higher chance of surviving over time as they develop trust and dependable food-sharing partnerships.

This kind of behavior is a direct result of the prediction made by game theory: cooperation is often the greatest long-term strategy for survival, even though betrayal may provide a short-term advantage.

Competition and natural selection

Obviously, cooperation isn't a factor in every evolutionary strategy. Another essential component of game theory and evolutionary biology is competition. There is perpetual competition between plants and animals for resources such as territory, food, and mates. Natural selection in these situations favors those who have the finest means of out-competing others.

For example, male deer engage in fierce battles using their antlers during mating season to attract females. Each deer must determine whether the chance of damage outweighs the reward of victory in these calculated risk standoffs. Since the gain of one deer equals the loss of another, this is a zero-sum game in game theory. Generation after generation, the stronger or more intelligent individual always prevails.

Evolutionary Stable Strategies (ESS)

Game theory also introduces the concept of an Evolutionary Stable Strategy (ESS), which is simply a strategy that, once embraced by a population, is difficult to replace with a different strategy. In other words, it's a strategy that works so well that it gets accepted as the standard.

The way birds settle territorial disputes is an example of an ESS (Evolutionarily Stable Strategy) found in the natural world. Many birds engage in what's known as a *display contest*, a ritualized competition in which they use non-violent displays such as fluffing their feathers, making loud sounds, or performing elaborate dances to intimidate rivals and establish dominance without actual fighting. This tactic, which helps avoid injury, eventually becomes the standard approach, as it benefits all parties involved by allowing them to survive and compete again in the future.

Nature's act of balancing

We can see that cooperation and competition are constantly balanced in nature, according to game theory. Species develop strategies based on what is most effective for their survival; male deer, for example, may compete furiously, whereas pack animals form alliances. The ongoing testing and improvement of these techniques in natural selection is comparable to how corporations evaluate their plans in highly competitive marketplaces.

Game theory helps us to understand these natural strategies and, thus, the complexity of life. It demonstrates that the rules of the game remain the same whether in a boardroom or the real world: using the information at hand to make the best decision possible, striking a balance between cooperation and competition, and adjusting as necessary.

The next time you witness a pride of lions cooperating to hunt or birds chirping in the trees, keep in mind that these animals are essentially engaging in a game of their own – one that has been honed over millions of years. And we now know more about how these amazing strategies work, all thanks to game theory.

PSYCHOLOGY AND COGNITIVE SCIENCE

Game theory isn't just about numbers, strategies, or logical decisions; it's also deeply connected to human behavior and psychology. The main goal of game theory is to comprehend how individuals make decisions, including what drives them and how encouragements or anxieties influence their actions. Through the convergence of game theory and psychology, we can acquire a deeper understanding of the reasons behind people's decisions in everyday life, personal relationships, and business.

Understanding human behavior through game theory

Basically, game theory is the study of behavior prediction. You can predict more accurately someone's actions and modify your approach if you are aware of their potential response to a certain circumstance. This is when psychology comes into play. Not all decisions made by people are motivated only by reason or logic. Various factors, including emotions, cognitive biases, past experiences, and social influences, can impact our attitude to a certain game or situation.

Consider, for instance, a negotiation. Theoretically, each party should rationally consider their options and make a decision based on what will benefit them the most. However, in real life, feelings such as pride, fear, or even guilt can influence decisions. By providing models that take into consideration how real people think and behave – rather than merely how they should – game theory helps in our ability to explain these psychological aspects.

Cognitive biases and decision making

The study of cognitive biases – those mental shortcuts that cause us to make illogical or inefficient decisions – is one area where game theory and psychology intersect. Whether we are aware of it or not, we are all biased, and this can have a significant influence on the decisions we make in every circumstance.

Consider the status quo bias, which occurs when people are more likely to stick with what they know than to take chances in the hopes of making a profit. According to game theory, this bias could explain why a business sticks with an antiquated approach when it is obvious that following a new market trend could produce superior outcomes. They hesitate to take risks because they are afraid of losing what they currently have.

Similar to this, loss aversion is a bias in which one experiences greater pain upon losing something than one does upon obtaining something equally valuable. Due to this bias, people often make too cautious decisions by putting loss avoidance ahead of gain pursuit. In competitive situations, in particular, game theory can be used to predict when loss aversion will manifest itself and how it will affect strategies.

The role of emotions

Another strong element that can change the course of any game is emotion. Whether you're playing a simple game of poker, arguing with a partner, or engaging in business negotiations, emotions have the ability to quickly take precedence over reason.

Think about anger, for example. Even when it's not in their best interests, someone who is angry may decide to punish the other player. According to game theory, this could result in illogical decisions that have negative effects on all parties. Game theorists can simulate these factors and provide ways to deal with emotionally charged situations by having a thorough understanding of how emotions such as anger, fear, or enthusiasm affect decision-making.

In a work environment, for instance, a team leader may become frustrated with a worker's performance and, in the heat of the moment, choose to remove them from a significant project. However, in the long run, this emotional response could be detrimental to the team's overall productivity. Even at times of intense emotion, game theory serves as a reminder to take a step back and think strategically, taking the wider picture into account rather than simply the immediate emotional reaction.

Game theory in group dynamics

The way individuals behave in groups is another aspect of psychology that game theory sheds light on. Group dynamics are complex and often entail power struggles, coalitions, and group decision-making. Game theory can help explain how these interactions take place in groups.

Consider social conformity, which occurs when individuals act following the crowd instead of thinking for themselves. People often behave this way in social or professional environments when they are hesitant to question the group's decision, even when they know it is not the right one. Game theory offers techniques to overcome group thinking by promoting varied opinions and more independent decision-making, and it can help predict when conformity will triumph over critical thinking.

Conversely, game theory can also emphasize how crucial group collaboration is. Through repeated interactions, people learn to trust one another and that cooperating often produces better long-term outcomes than acting selfishly. Here's where well-known game theory models, such as the iterated prisoner's dilemma, come into play; they demonstrate how persistent cooperation can develop into a reliable and beneficial strategy in group situations.

Practical application of psychology in game theory

So, how do you use your knowledge of human psychology in your daily life? First of all, acknowledge that human decisions are rarely entirely logical. When engaging in any kind of interaction with other people, be it team management, business negotiation, or even family vacation planning, take time to think about the psychological aspects involved. Does the individual you're interacting with fear losing something? Do they have an emotional response, or are they influenced by group dynamics? You can also modify your approach to better handle the circumstance by taking different things into account: for example, if you are aware of someone's propensity to avoid risk, you could present your idea in a way that highlights stability and security to ease their concerns.

Although game theory provides an organized method for examining these dynamics, its human component is what's really fascinating. The next time you find yourself facing a difficult decision, keep in mind that understanding the underlying motives behind the decisions made around you is just as important as applying logic. You'll be more capable of navigating the game if you do this in relationships, career, or life in general.

ECONOMICS AND POLITICAL SCIENCE

Game theory is an important tool for understanding political science and economics because it lets us simulate and predict how individuals, organizations, and governments will act in cooperative or competitive environments. Game theory provides a framework to look past individual behaviors and examine the bigger strategies at play in a variety of contexts, including markets, trade agreements, elections, and international diplomacy.

Game theory in economics

Game theory offers a framework in economics for examining how businesses, customers, and even entire industries make decisions that affect other players in the market as well as themselves. Understanding competition is one of the most popular uses of game theory in economics.

Imagine two companies attempting to determine the costs for a similar product. A company may see an increase in business if it lowers its pricing, yet if both drop their prices, profitability may suffer. Economists may simulate similar situations using game theory to examine how different firms would respond to one another. Consider applying the prisoner's dilemma, for instance, to Coca-Cola and Pepsi, two rival businesses. Both businesses could marginally grow their market shares if they spent a lot of money on advertising, but it would be very expensive for them both. On the other hand, they could both save money if they cut back on their advertising. Game theory clarifies why businesses could decide to openly compete with one another or to make an unspoken agreement (without breaching the law) to limit the power of their competitors in order to maximize profits.

Game theory is also often applied in oligopolies, in which a small number of enterprises control the majority of the market. Businesses must take their competitors' actions into account when making decisions in certain circumstances, in addition to their own best interests. Based on game theory, economic models illustrate how companies might work together (even covertly) to maintain high pricing or compete fiercely to dominate a market.

Game theory in political science

Game theory in political science sheds light on the decision-making processes of governments, political parties, and leaders – often in the face of uncertainty or competition. Consider elections as an example. Political parties are not only responding to people, but they are also monitoring the actions of their competitors. Every action, like coming up with a new policy, giving a speech in public, or joining an interest group, is a part of a bigger scheme.

Voting systems are one well-known instance in politics. Candidates build their platforms to gather the most votes, but rather than remaining true to their initial beliefs, candidates may soften their stances in order to win over more supporters. This is particularly true in two-party systems where winning depends on a candidate's ability to reach people in the 'middle'. Political candidates often soften their positions as an election approaches because they are trying to secure the most votes possible through calculated strategies.

International relations is another field in which game theory is very important. Countries must take into account how other nations will react in addition to what is beneficial for themselves while negotiating trade agreements, arms sales, or even peace accords. Models based on game theory help policymakers in predicting various outcomes and identifying the best course of action. As mentioned previously, game theory was applied, for example, to nuclear deterrent measures during the Cold War, specifically during the Cuban missile crisis.

Real-world example: trade wars

The U.S.-China trade conflict serves as a practical illustration of how game theory can be applied to political and economic strategies. To obtain power in trade discussions, both nations, in this case, applied taxes on each other's goods. "How will each country react to the other's actions?" is the question that game theory helps to answer in this situation.

China may retaliate with taxes of its own if the United States raises tariffs, which would be detrimental to both nations. However, one nation may lose face or negotiation leverage if it gives in. These kinds of scenarios are modeled by game theory in order to illustrate potential outcomes and pinpoint equilibrium points – the moments at which both nations agree to cease escalating and reach a compromise.

Game theory helps to describe this situation's delicate risk-reward balance. Every move is a calculated risk, with both countries attempting to safeguard their interests while projecting the potential response of their competitor.

It is important to understand how game theory shapes political and economic strategy because it influences decisions that define the world around you. The concepts of game theory are applied to everything from retail prices to political outcomes and global policy. Understanding these trends can help you predict leaders' next actions and understand why organizations, governments, and individuals make the decisions that they do.

PART 4

PRACTICAL APPLICATIONS AND CASE STUDIES

CHAPTER TEN
IMPROVING NEGOTIATION SKILLS

Negotiation is a game that has many and varied applications in supply networks, international wars, boardrooms, families, and teams, at its foundation. People engage in various negotiation games in different real-world scenarios, such as making a purchase in a store, haggling for a new car, convincing kids to finish their schoolwork, or planning a vacation with a partner.

Exchanges of any kind are a necessary component of the process of negotiation, which is when two or more parties discuss or work together to come to a decision or agreement. As such, when people engage in negotiation, a variety of forces and dynamics may surface inside each party. Undeniably, a better understanding of these factors will lead to better performance, which is why game theory is occasionally used in negotiations to maximize positive results.

A set of about thirty games that simulate various real-world situations form the basis of game theory. Four are very important in terms of negotiations. These are the 'stag hunt', the prisoner's dilemma, and the 'trust game'.

A game is never always right or wrong. The best decision will depend on the details and might even need to change as the negotiation progresses. The important thing to remember is that we can choose the game we negotiate. Even while we might not be able to influence the games the other side chooses, knowing game theory can help us figure out their approach and tip the odds in our favor.

The 'trust game'

The proposer in a 'trust game' can decide how the parties will split the benefits as a result of their position of power over the other (the responder).

Given their full helplessness and lack of actual power in this situation, the respondent decides to start by offering a gift or concession in the hopes that it may sway the decision in their favor. Giving a gift is an act of judgment by the weaker party with the expectation that it will lead to a better outcome and that it will be repaid.

Trust can work either way when one party is more powerful than the other; the weaker party, however, has the option to make a move in this if they choose to give a gift. Without it, the only role in the game is that of a dictator (the proposer), who has complete control over whether or not to give in to the requests.

The gift can actually take on a variety of shapes. For instance, to attract or retain valuable clients, a supplier may offer incentives such as loyalty bonuses, complimentary services, or agree to less favorable terms for a strong buyer. In return, the buyer might suggest that the supplier use them as a reference, promote the supplier to their own customers, or provide flexibility regarding timelines.

The prisoner's dilemma

The most well-known game, the prisoner's dilemma, consists of two players getting together, deciding on a plan of action or a bargain, and then splitting off to carry out the agreement, with each player choosing whether to follow through on the agreement or not.

The term originates from an account written by Albert Tucker from Stanford University, which was published in the Philadelphia Inquirer. Tucker described how two burglars were apprehended by the police, placed in separate cars, put in different interview rooms, and thoroughly interrogated in an effort to get a confession. Each criminal was told separately that they needed to think twice before confessing and bringing the other person into it.

The authorities can only charge both with lesser offenses if neither confesses, in which case they will each spend a year in jail. Each burglar will serve a ten-year term if they both confess and incriminate the other. On the other hand, if one confesses and implicates the other but the other does not, the person who assisted the police will be released from prison, and his accomplice will be sentenced to 10 years.

The decision one takes determines the prisoner's dilemma's results (benefits or penalties), but it's important to remember that if the other party doesn't try the same thing, one can gain more by tricking the other. The fundamental principle of the game is self-interest, which motivates acts that seem reasonable to the player in the given circumstance. But when self-interest takes priority, both sides lose out in the end.

When fish stocks declined in the 1970s, fishing quotas were implemented: they serve as an excellent illustration of the prisoner's dilemma. Since individual fishermen were more concerned with maximizing their income than with maintaining a long-term position, quota enforcement proved to be challenging. In areas where two nations share fishing grounds and both guarantee that fishermen adhere to the rules, the catch would be modest yet sustainable. But the catch would be abundant if one nation ignored fishermen who took more than what was agreed upon. Although there would be fewer fish the following year, this would improve the economy. If both nations took this action, they would benefit initially, but as the fish ran out, they would have to look for new places to fish. Currently, there are about 250 fishing quota systems in existence worldwide, and they are enforced by monitoring systems and satellite tracking, which help ensure that fishing limits are respected across international waters.

The stag hunt

The stag hunt describes a contradiction between social cooperation and going with the safe route.

Two people set out on a hunt. Without knowing the other's decision, each person is free to decide whether to search for hares or stags. An individual can hunt a hare on their own, but a hare is less valuable than a stag. But he will definitely eat if he goes hare hunting. To be successful in stag hunting, a person needs the collaboration of their partner. The likelihood of a successful stag hunt increases significantly when there is more than one hunter.

The advantage that working together brings to both sides is crucial. The main difference with the prisoner's dilemma is that there are two states of mutual benefit: if both sides hunt hare, they both gain since they can eat and if they both hunt stag, they both gain more.

Choosing which game to play

Game theory may be applied to various situations, from global peace to sustainable fishing. However, in any negotiating situation, it is critical to determine which game has the greatest chance of producing the desired result.

Even while game theory can be used to explain several facets of human nature and behavior, it is not commonly included in company development programs. Usually, sales teams don't choose to play games like the prisoner's dilemma; they just happen to end up doing so.

Planning games ahead of time is a good idea when it comes to negotiation strategies. Depending on the situation, it may be acceptable to play a sequence of games and alternate between them. We must always be conscious of the current situation and mindful that things can change at any time.

NEGOTIATIONS IN BOTH PROFESSIONAL AND PERSONAL CONTEXTS

Whether you're trying to resolve a conflict with a loved one or discussing a commercial deal, negotiating can feel like a high-stakes game. With the help of game theory, you may effectively navigate these circumstances and gain the knowledge necessary to understand the dynamics at play and make more informed judgments.

Leverage the power of commitment

In game theory, commitment strategies can significantly influence the direction and outcome of a game. Commitment involves signaling clearly to the other party that you are serious about your position and prepared to follow through. When used thoughtfully, this approach can strengthen your stance and compel others to respond accordingly.

In a business negotiation, for instance, you can establish commitment by setting clear terms or deadlines. Saying, "We need a decision by the end of the month, or we'll have to explore other options," sends a decisive message that pressures the other side to act without unnecessary delays.

In personal interactions, commitment can be less overt but equally impactful. Imagine discussing shared responsibilities with a partner. By clearly committing to an agreed role, like saying "I'll handle the cooking if you take care of the cleaning," and consistently following through, you reinforce trust and clarity in the arrangement. This steady commitment establishes a foundation for cooperation and mutual respect, ensuring that each party remains accountable to their part of the agreement.

Use the tit-for-tat strategy

Tit-for-tat is a well-known game theory strategy in which you match the opposing player's moves with moves of your own. You cooperate if they cooperate. You respond to their competitive or hostile behavior by matching it, demonstrating to them your unwavering resolve.

This approach has the potential to be very successful in negotiations with professionals. Make concessions or meet halfway if the other party is reasonable and makes a fair offer in return. However, do not hesitate to respond aggressively if they begin to play hardball. This is where balance is important: remain firm yet fair. When one party recognizes that working together is in their best interests, it often fosters a more cooperative environment.

Setting boundaries in personal relationships can benefit from the tit-for-tat method. You reciprocate a friend's or partner's respect for your boundaries when they do. However, you respond forcefully if they go too far, letting them know that you value justice and decency. This creates healthier behavioral and communicative patterns over time.

Know when to walk away

Ultimately, one of the most crucial things to remember from game theory is when to give up. Sometimes, it's best to just walk away from a discussion altogether. This could be the result of a lack of cooperation from the other party or the decision that the benefit isn't worth the trouble.

Walking away might provide you influence in a commercial setting if a deal is no longer possible for any reason (whether the supplier is demanding too much or the terms are becoming unfavorable). Being prepared to back down might cause the other side to reevaluate their position, which often results in improved offers or fresh conversations.

In interpersonal contexts, this could be as simple as calling it quits on an argument or choosing not to participate in a pointless debate. There are situations when it is more expensive – in terms of stress, time, or emotional energy – to continue a negotiation than it is beneficial. Effective negotiation skills are only as vital as knowing when to back off.

APPLICATION EXERCISE

Let's put the theory into action! These exercises will help you practice game theory negotiation strategies and gain more confidence in real-life situations. Remember, the goal is to sharpen your skills so that you can approach negotiations – whether at work or in your personal life – with clarity and strategy. Let's dive in!

Exercise 1: Role-playing a salary negotiation

Let's say you're negotiating a raise with your boss. Here's the scenario: you've been with the company for two years, consistently meeting your targets, and you feel it's time for a salary increase. Your boss, however, has a limited budget and may be hesitant to approve the raise.

Your task: write down your strategy before entering the negotiation. Consider the following:

- What are your payoffs (e.g., salary increase, additional benefits, professional growth)?
- What might your boss's payoffs be (e.g., staying within budget, keeping you happy as a valued employee)?

- How can you cooperate instead of competing (e.g., offering to take on more responsibility, suggesting a raise over time)?

Once you've outlined your strategy, rehearse the negotiation out loud or with a friend. Afterward, reflect on how it felt and whether you were able to incorporate game theory principles.

Exercise 2: Practicing tit-for-tat strategies in personal conflicts

Think about a recent disagreement you've had with a friend or partner. Maybe it was about household responsibilities, where both of you held firm positions.

Your task: Reflect on how you could have applied the tit-for-tat strategy. Answer the following questions:

- How could you have responded if they cooperated (e.g., showing appreciation, making a small compromise)?
- If they acted competitively or didn't respect your boundaries, how could you have matched that behavior while still staying fair?

Now, imagine the conversation again, but this time with tit-for-tat in mind. How does this change your approach? Write down any thoughts that cross your mind about how you might handle a similar situation in the future.

Exercise 3: Simulating a business partnership negotiation

You're the founder of a startup looking to partner with another company. Both of you offer complementary services, but you need to negotiate terms on how to split profits.

Your task: Using the payoff matrix concept, create a simple chart with different possible outcomes of the negotiation. Think of what you and the other company might gain or lose based on different decisions, such as:

- You offer 50/50 profit sharing
- They insist on 60/40 in their favor
- You propose joint marketing initiatives

After creating the payoff matrix, consider how you can frame the conversation to highlight a cooperative strategy where both companies benefit. How would you present this idea to the other party?

Exercise 4: Negotiating without full information

Here's a scenario where you don't have all the facts upfront: you're buying a used car, but you're unsure if the seller is completely transparent about its condition.

Your task: Think about how you can use probability in your decision-making process. For example:

- How likely is it that the car has an undisclosed problem?
- What signals or behaviors from the seller would increase or decrease that probability?

Write down your negotiation strategy, including a series of questions you could ask to gather more information and how you might use this information to adjust your offer. Practice negotiating this scenario with a friend and see how your game theory-based thinking plays out.

Exercise 5: Walking away with confidence

Knowing when to give up a negotiation can sometimes be the best course of action. Think of a situation where the negotiation might not be in your favor, such as a job offer that doesn't meet your needs or a business deal with too many risks.

Your task: Pick a real or hypothetical scenario where you feel like you might have to walk away. Answer the following questions:

- What would be your point of no return (e.g., the minimum salary, the terms of the deal)?
- What signals would tell you it's time to walk away?

Write down the strategy you would use to communicate your decision professionally and assertively. Then, practice how you'd say it out loud. Walking away is not about being confrontational – it's about being clear on what you value and sticking to it.

Exercise 6: Resolving a personal dilemma with cooperation

Let's bring this closer to home. Think of a time when you and a family member or friend disagreed on something important, like where to go for a vacation or how to split time for family gatherings.

Your task: Using cooperation as the focus, write down how both of you could have won in this situation. How could you frame the conversation so that you're both aiming for mutual benefits rather than a win-lose outcome?

Reflect on how you can approach future disagreements with this mindset. After you've written down your thoughts, try discussing a similar topic with that person and see how using cooperation changes the tone of the conversation.

CHAPTER ELEVEN
REAL-WORLD CASE STUDIES

Game theory, a mathematical model of cooperation and conflict, has been applied to many real-world situations. It offers a framework for understanding and predicting outcomes in a variety of contexts, including social interactions and economic discussions. In this chapter, we'll look at some case studies coming from real situations.

FAMOUS BUSINESS DISPUTES

Let's examine a few well-known business disputes for which game theory proved to be a crucial tool in finding solutions. These real-world cases will demonstrate the effectiveness of game theory in business conflict resolution and strategic decision-making.

1. Pricing war: American Airlines vs. Delta

Consider a scenario with a pricing war between two significant airlines: American Airlines and Delta. They are both lowering the cost of their tickets aiming to draw in more customers. The situation is a typical example of a prisoner's dilemma, with each airline trying to outbid the other at any cost.

However, this ruthless competition cannot continue. Both airlines end up losing money, and reduced services could mean a worse overall experience for customers. They require a calculated approach that goes beyond simply responding to one another's actions in order to end this loop.

In theory, game theory suggests that a cooperative approach could be the best course of action in this situation. For example, Delta and American Airlines could consider working together through a code-sharing arrangement or coordinated flight scheduling. Although this is a proposed strategy rather than an actual event, such calculated collaboration could increase profitability and enhance passenger services for both airlines. By avoiding damaging price wars, this hypothetical solution illustrates a potential win-win scenario.

2. The Cola Wars: Coke vs. Pepsi

Another interesting example is the Coca-Cola versus Pepsi-Cola war. These two titans have been engaged in a decades-long battle for supremacy in the soft drink industry. Their competition isn't limited to who can provide superior taste; it also involves distribution, marketing, and sometimes even a little-known formula adjustment.

The idea of strategic maneuvers from game theory is one of the battle's main strategies. Both companies usually employ a combination of strategies in their pricing plans, sponsorship agreements, and advertising campaigns. For example, Pepsi may counter-campaign or offer a special discount in response to Coca-Cola's major marketing campaign. The goal is to anticipate one another's movements and skillfully counter them.

The 'Pepsi Challenge,' in which Pepsi offered blind taste tests to demonstrate that people preferred their drink over Coca-Cola, was a turning point. This was a calculated action to change public opinion using game theory. To preserve its market dominance, Coca-Cola had to adjust by improving its marketing strategies and introducing new products.

3. The Microsoft antitrust case

One well-known legal dispute where game theory was heavily employed was the Microsoft antitrust lawsuit. Microsoft was accused by the US government of engaging in monopolistic activities, specifically concerning the way it bundled its Internet Explorer browser with the Windows operating system. They claimed that this hurt consumers and inhibited competition.

Game theory helps to better understand the dynamics of monopolies and competitive practices. The bundle offered by Microsoft was similar to a dominant strategy in which the business took advantage of its position to obtain an unfair advantage over the competition. The concept of Nash equilibrium in game theory helps analyze how a company's pricing and production decisions impact the market and its competitors, as it identifies points where neither the monopolist nor any potential competitors would benefit from changing their strategy unilaterally.

In order to resolve the lawsuit, Microsoft had to consent to a number of modifications to its business procedures, one of which was granting customers the option to select different web browsers. The goal of this settlement was to bring back the competitive equilibrium in the software industry through a calculated compromise. The case demonstrated how market competition and antitrust issues can be examined and addressed using game theory.

4. The Apple and Samsung patent dispute

Another interesting example is the ongoing legal battle between Apple and Samsung on smartphone patents. Both businesses accused one another of infringing on the other's patents, which sparked a series of high-profile cases around the world.

Game theory can be applied here to examine dispute strategies and bargaining techniques. Both businesses sought to obtain favorable decisions and settlements to increase their payoffs. For example, Samsung wanted to demonstrate that Apple's patents were invalid, whereas Apple's strategy comprised claiming that Samsung's designs were too close to their own.

Ultimately, a settlement was reached by both parties, which included agreements on future patent use as well as monetary compensation. This resolution serves as an example of how game theory concepts can direct the negotiation process in complex legal disputes, helping businesses reach amicable resolutions and averting protracted confrontations.

This shows how game theory is a tool that businesses use to manage conflicts, decide on strategies, and settle disagreements. Knowing the underlying strategic relationships can help with improving outcomes and offer useful insights into various forms of conflict resolution, such as pricing wars, competitive strategies, and legal disputes. So, keep in mind that game theory might hold the key to finding a workable solution the next time you find yourself in a challenging business scenario.

HISTORICAL EVENTS

Game theory has practical implications that have influenced some of the most important historical events. By understanding these applications, you can observe how game theory ideas and strategic thinking might impact decisions in high-stakes scenarios.

1. The prisoner's dilemma in Cold War espionage

Espionage and intelligence gathering were essential aspects of international affairs throughout the Cold War. The idea of the prisoner's dilemma was especially relevant here. Imagine that two spies, one from each side, are apprehended and subjected to separate interrogations. Each must choose between keeping quiet and betraying their nation by disclosing secrets. Both receive a reduced sentence if they stay silent. If one speaks up and the other does not, the one who speaks out is punished severely, while the one who betrays gets away with it. Should they both turn traitors, the penalty will be mild. The main problem is determining whether to trust the other spy to keep quiet or to turn on them in exchange for a perhaps better result.

According to game theory, in situations like this, reciprocal cooperation – that is, both spies remaining silent – leads to the greatest possible collective outcome. However, people often turn to a self-preservation approach due to a lack of trust and the possibility of betrayal. This dynamic affected how nations conducted their intelligence operations and their connections with allies in numerous espionage scenarios.

2. The Berlin airlift: strategic cooperation

The Berlin Airlift of 1948–1949 is another historical event where game theory techniques were used. Berlin was divided into East and West after World War II, with Soviet-controlled East Germany encircling the city's western area. The Allies had to choose between abandoning West Berlin or finding a means of supplying it by air after the Soviet Union blockaded all ground access to the city.

Choosing the second option, the Allies organized a huge airlift to deliver supplies to the residents of West Berlin. Game theory here helps us understand the strategic choices that must be made. The risk of escalation and the necessity of upholding their pledge to the Berlin population had to be balanced by the Allies. On the other side, the Soviets had to choose between negotiating and maintaining their embargo. The airlift served as an example of the strength of strategic commitment and teamwork. The Soviet Union took a risk by deciding to maintain the embargo while the Allies delivered supplies, which may have led to more diplomatic humiliation and an intensification of the conflict. The blockade was eventually removed by the Soviet Union, and the sustained supply operations of the Allies were effective.

3. The Treaty of Versailles and post-war negotiations

Game theory played a vital role in the crucial diplomatic negotiations that led to the Treaty of Versailles after World War I. In order to maintain peace and further their own national interests, the three main powers – France, Britain, and the United States – had to negotiate agreements.

We can understand the dynamics of negotiations and the tactical choices made by each side by using game theory. While Germany aimed to avoid harsh punishments, the Allies intended to guarantee that Germany could never wage war again. In hindsight, the ensuing treaty contributed to Germany's political and economic instability by imposing significant territory losses and costly reparations on the nation.

Another good example of how game theory may help understand the complex dynamics of post-war diplomacy and conflict resolution is the relationship between national interests and strategic negotiation. The Treaty of Versailles serves as a lesson in balancing competing interests and the long-term costs of negotiation strategies.

KEY LESSONS FROM THE CASE STUDIES

When we reflect on these historical case studies, we can learn several important lessons about the use of game theory in high-stakes disputes and resolutions. These insights can be extremely helpful when managing a personal relationship, navigating a complex business deal, or making strategic judgments. Here are the main key points to keep in mind.

1. Anticipate and understand your opponent's moves

The importance of anticipating the responses of your opponent is among the most crucial lessons to be learned from the Cuban Missile Crisis. It was incumbent for President Kennedy and Premier Khrushchev to predict each other's response to their respective acts. Any negotiation or confrontation needs this kind of strategic foresight. When you're making a decision, take into account the opinions of others around you. Think about their objectives, worries, and possible reactions. Gaining insight into their viewpoint can enable you to make better, more sensible decisions.

2. Look for win-win resolutions

We can also learn from reaching mutually beneficial agreements from the Cuban Missile Crisis. Kennedy and Khrushchev sought a solution that would allay tensions and solve their individual concerns rather than just concentrating on their own interests. When negotiating or resolving disputes with others, try to find solutions that work for everyone. Seek for situations where everyone can benefit and come away with something. This strategy results in more enduring agreements as well as improved relationships.

3. Develop cooperation and trust

Cooperation and trust are important, as demonstrated during the Berlin airlift. To overcome a formidable obstacle, Berlin's residents and the Allies had to rely on one another. For any discussion or partnership to be successful, trust is an essential component. Develop trust by being dependable, being open and honest, and keeping your promises. It is usually more beneficial to cooperate than to argue, so make an effort to cultivate solid, reliable bonds.

4. Get ready for multiple scenarios

The Allies in the Berlin Airlift had to prepare for a variety of scenarios. Game theory helps us to avoid concentrating on just one strategy and instead trains us to plan for a variety of scenarios and outcomes. Make backup plans and take into account a variety of possible outcomes while making decisions. In this manner, you'll be able to adjust to changing conditions and deal with unforeseen events skillfully.

5. Control your emotional reactions

We learn about the importance of emotions in decision-making through the prisoner's dilemma and the espionage scenario. Emotions such as the desire for self-preservation and fear of betrayal might affect our decisions. Recognize how emotions influence your decisions and make an effort to regulate them. You can make better choices and have greater mental clarity in challenging situations if you possess emotional intelligence.

6. Balance your short-term and long-term goals

The Treaty of Versailles serves as an example of what happens when short-term goals take priority over long-term stability. Although the treaty was intended to avert future hostilities, it also caused Germany to become unstable over time. When making decisions, consider both the short-term and long-term effects. Think about how your decisions will impact both the present and the future. Make decisions that will help ensure stability and prosperity in the long run.

7. Apply strategic thinking in everyday situations

Finally, keep in mind that game theory applies to more than just business and international disputes. Everyday circumstances can also benefit from the use of strategic thinking concepts. Use these lessons to direct your approach when managing a team, settling a family dispute, or negotiating a contract. Consider all options, plan ahead, and look for solutions that will satisfy all parties.

CHAPTER TWELVE
IMMEDIATE APPLICATIONS

You may believe that only economists, mathematicians, or theorists can truly understand the abstract concept of game theory. The truth is that the principle of game theory permeates every aspect of our daily existence. We make decisions every day that require some level of planning and thought for potential reactions from other people.

APPLYING GAME THEORY IN DAILY LIFE

You may be surprised by the numerous ways game theory is applied in daily life. It helps us make better decisions by taking into account not only what we desire but also what others might do. This applies to anything from personal financial decisions to navigating difficult social circumstances.

1. Deciding on investments

Game theory pushes you to consider the market strategically when making financial investments. Rather than simply following the whims of the moment, take into account the actions of other investors and how they may impact the market. For instance, some investors may sell off equities during uncertain economic times, which might lower prices. According to game theory, you may profit from this by purchasing cheap stocks and holding onto the belief that their prices will increase until the market stabilizes. By planning a few steps ahead, you can avoid making impulse decisions and instead use a long-term perspective.

2. Resolving personal conflicts

We have disagreements in our interpersonal interactions, whether they be with a partner, friend, or relative. Game theory educates us to take into account not just our own viewpoint but also the motives and possible responses of others. Let's imagine you and your partner are at odds over what to do this weekend. Consider your partner's ideals and their potential reactions to such compromises before becoming really aggressive. You have a better chance of coming to a calm agreement that benefits you both if you propose a solution that gets them what they want and what you need. Game theory promotes cooperation over confrontation, which results in more win-win situations.

3. Splitting the bill with friends

Deciding how to split the bill at a restaurant can occasionally be tricky, especially if one person orders more than the others. Here again, game theory can be useful. It could make sense to order similar items to prevent feeling like you're overpaying when you go out with friends, and the group usually shares the tab equally. But game theory proposes talking about an alternate strategy, such as each person paying for what they ordered, if you find someone continually ordering significantly more while everyone else is trying to save money. In this manner, unneeded stress is avoided, and everyone believes that the split is fair.

4. Negotiating a pay raise

The core idea of game theory is strategic positioning, which includes wage negotiations. When getting ready to ask for a raise, it's critical to consider your employer's position in addition to your own desires. What could they possibly offer? What are the financial limitations of the company? You're more likely to come to an agreement if you make your case in a way that benefits both you and your employer. Examples of this include showing how your work has benefited the firm financially or helped it grow. Making it seem like both parties are benefiting is the aim of creating a win-win scenario.

5. Deciding where to go on vacation

You can even use game theory to plan group vacations. When choosing a destination for a trip with friends or family, there may be a deadlock as each person pushes for their own choice. Think about making a calculated compromise rather than forcing your decision. You may suggest a rotation in which someone's preference is selected this time, but someone else's will be the next time. In this manner, nobody feels forgotten, and everyone feels heard. By demonstrating your consideration for equity and future cooperation, you can encourage others to be more amenable to compromise.

6. Using limited information to make daily decisions

Game theory may also be useful in situations when you do not have access to all the facts. Imagine you're shopping for a car, but you're not sure if the price you're getting is the best available. Game theory advises you to obtain as much information as you can in this situation (for example, by using online resources or comparing pricing at several dealerships). Then, when engaging in negotiations, take into account the salesperson's potential mindset – what is their bottom line? You can make a better, knowledgeable, and confident choice by exercising strategic thought and taking into account the viewpoint of the opposing party.

7. Effective time management

Time management is another real-world application of game theory. Consider your time as a resource, and each decision you make about how to use it is a 'game' with a potential outcome. When trying to strike a balance between your personal life, career, and interests, game theory suggests giving priority to the pursuits that will yield the most long-term gains. While spending too much time on distractions can reduce productivity, perhaps investing time to master a new skill can pay off in the long run. By thinking through potential outcomes and strategically managing your time, you may make better decisions that advance your long-term goals.

QUICK EXERCISES

With these quick exercises, let's put some of the game theory concepts we've covered into practice. These will enable you to see how game theory can improve your decision-making and will help you begin to think strategically in everyday circumstances. All you need is some reflection and practice!

1. Play the investment game

Let's say you have $1,000 to put in bonds or equities. Investing in equities carries a greater chance of financial loss as well as greater profit potential. Bonds have lower returns but are safer. Now consider what the overall market may be doing: are investors buying equities, which carry risk, or are they swarming to safe havens like bonds? Which strategic decision would you make to optimize your return based on what others might do?

2. Win-win negotiation practice

Imagine yourself going into a little negotiation, such as asking for a discount, talking to your partner about a household chore, or asking a coworker for assistance. Outline the potential benefits for each party before the talk. What can you provide that the other person will also benefit from? You'll increase your chances of success by phrasing your request in a way that demonstrates mutual benefit.

3. Time management technique

Examine your weekly agenda for a moment. List the things you need to do and order them according to importance in the long run against immediate gratification. Now, apply game theory by thinking of your time as a limited resource. Which task will best position you for rewards down the road? Make decisions based on your short-term and long-term goals; picture it as a game where you're trying to maximize your moves to win.

4. Social media strategy

Select a social networking platform and consider the various ways users interact with posts, such as likes, shares, and comments. Now, consider your options strategically: what would people value most from a post if you wanted to boost engagement on it? What type of content usually receives the greatest engagement, and how can you use that information in your own posts? This will help you in considering the reciprocal effects of people's behaviors.

These quick exercises are meant to help you in putting game theory into practice. As you work through them, you'll begin to see patterns in both your own and other people's decisions.

CUSTOMIZABLE TOOLKIT

Finally, we provide you with a customizable toolkit to enable you to apply game theory to both your personal and professional life. Whether you're trying to manage team dynamics, navigate complex negotiations, or make better decisions in your day-to-day life, this toolkit offers a useful template that you can readily customize to your circumstances.

1. Decision-making template

Sometimes, laying out your options visually makes all the difference. Here's a simple decision-making template to guide you through tough choices:

Step 1: List your options. Which steps could you take to accomplish this?

Step 2: Identify the players. Who else is involved, and how do their decisions impact your outcome?

Step 3: Assess the payoffs. For each option, what's the potential reward? Consider both short-term and long-term payoffs.

Step 4: Consider potential risks. What are the risks associated with each choice? How much risk are you comfortable taking?

This template will help you break down your choices into manageable steps, giving you clarity before making your move.

2. Strategy chart for negotiations

In negotiations, having a clear strategy is vital to success. Use this strategy chart to map out your next negotiation, whether it's with a supplier, a client, or even a personal agreement at home:

Factor	your position	their Position	possible compromise
Desired outcome	What do you want?	What do they want?	What middle ground exists?
Leverage	What power do you hold?	What power do they hold?	How can you balance it?
Key Arguments	What are your strongest points?	What are their strongest points?	How do you counter their points?
Best alternative?	What is your backup plan?	What is their backup plan?	What happens if you walk away?

This chart helps you focus on what both sides bring and highlights areas where compromise is possible, making your negotiations more effective.

3. Self-assessment: trust and cooperation

Use this self-assessment to evaluate how trust and cooperation play a role in your personal or professional relationships.

Question 1: How comfortable are you relying on others in essential decisions?

Question 2: How often do you communicate your goals and expectations to the people involved?

Question 3: When conflict arises, do you seek win-win solutions or focus on winning the argument?

Question 4: How much time and effort do you invest in building trust with others?

By answering these questions, you'll better understand how game theory can help you foster cooperation and trust in your relationships. Your chances of success will increase with the amount of trust you cultivate.

4. Payoff matrix for team projects

Understanding how each team member's actions affect overall success is crucial if you're leading a team. This payoff matrix will help you map out possible outcomes for different decisions.

Team member action	Positive outcome for team	Negative outcome for team
Member A works independently.	Increased productivity	Lack of collaboration
Member B collaborates with others.	Greater innovation	Potential disagreements
Member C delays work.	Decreased team morale	Missed deadlines

This simple matrix will help you visualize the ripple effects of individual actions within your team, helping you better manage group dynamics and decision-making.

5. Conflict resolution worksheet

When a conflict arises – whether in a work setting or a personal one – use this worksheet to approach it strategically.

Step 1: Define the conflict.

Step 2: Identify each party's goals and motivations.

Step 3: List potential solutions that benefit both sides (cooperative strategies).

Step 4: Consider what would happen if no solution is reached (the worst-case scenario).

Step 5: Take action by selecting the most balanced, fair solution.

This worksheet will give you a clear roadmap to resolve conflicts in a way that preserves relationships and leads to positive outcomes for everyone involved.

This toolkit is designed to be flexible, so feel free to adapt these templates and exercises to your unique situations. Whether you're handling negotiations, building trust, or managing conflicts, having a strategic game plan can make all the difference. Think of this toolkit as your guide to applying game theory in everyday life.

CONCLUSION

As we reach the end of this book, let's take a moment to reflect on the powerful insights you've gained. Throughout this journey, we've explored how game theory is more than just an academic concept – it's a practical tool you can apply to basically every aspect of your life. Whether negotiating a better deal, managing relationships, navigating complex business strategies, or making everyday decisions, you now have the frameworks and strategies to approach these challenges confidently.

We've covered the essentials of game theory, from understanding players' roles, strategies, payoffs, and equilibrium to diving deep into real-world applications in areas like business, politics, and personal relationships. You've seen how decision trees, payoff matrices, and trust dynamics shape outcomes and how emotions and risk-taking influence our choices. Most importantly, you've learned that game theory isn't about predicting the future – it's about giving you the tools to anticipate outcomes, navigate conflicts, and create win-win scenarios.

Now, it's your turn to put everything you've learned into practice. The beauty of game theory lies in its versatility: it's not just for economists or mathematicians, it's for anyone who wants to make smarter, more strategic decisions. Use this knowledge to analyze your next business deal, improve relationships, or tackle daily challenges. Each decision you make is an opportunity to leverage the principles of game theory to achieve better outcomes.

Consider this the start of your road to becoming an expert decision-maker and determining your own future. These strategies will become more second nature to you the more you use them. Game theory can be your secret weapon, empowering you to think strategically, act decisively, and ultimately achieve the results you've been aiming for in both your professional and personal life.

The world of decision-making is complex, but with game theory in your toolkit, you can turn complexity into opportunity. Go out there and start mastering your decisions – success is just a few intelligent moves away.

EXCLUSIVE CONTENTS:

Thank you for choosing *Game Theory Mastery*! To best support you on your journey toward more effective decision-making and a deep understanding of game theory, we have included exclusive content that only readers of this book can access.

1° Game Theory Mastery Audiobook:

Listen to *Game Theory Mastery* on the go! This audiobook allows you to absorb the key concepts of game theory wherever you are, ensuring that your learning never stops. Perfect for those who prefer active listening, our audiobook is a complete and convenient resource for fully immersing yourself in the subject.

2° Strategic Decision Making Guide:

Access a practical, structured guide to refine your decision-making process. The *Strategic Decision Making Guide* offers specific, applicable tools for making strategic decisions in both work and life. From the basics of rational choices to advanced methods, this complementary manual is ideal for those who want to apply the principles learned in the book.

3° Practical Exercises with Real-World Scenarios:

Test your understanding with a series of exercises designed to simulate real-life decision-making scenarios. This section allows you to directly apply game theory concepts to practical, challenging situations, helping you strengthen your skills and prepare for complex strategic decisions.

4° Emotional Intelligence and Communication Book:

Boost your interpersonal skills with the *Emotional Intelligence and Communication Book*. This content provides insights into understanding emotions, improving communication, and fostering better relationships. It complements your game theory journey by enhancing your ability to navigate complex social interactions and achieve win-win outcomes in both personal and professional settings.

Each of these exclusive contents is designed to enrich your reading experience and provide you with a multidimensional approach to game theory. We recommend combining reading, listening, and practice to maximize the benefits of your journey.

To download the exclusive contents, scan the QR-Code:

I sincerely hope you enjoyed

Game Theory Mastery!

If you did, consider leaving a review.

Your feedback not only helps other readers but also allows me to continue improving and providing valuable content.

A small gesture that can make

a big difference!

REFERENCES

Admin. (2019, October 31). *Game Theory and its Applications - Canadian Institute For Knowledge Development.* Canadian Institute for Knowledge Development. https://cikd.ca/2019/10/31/game-theory-and-its-applications/

Andreadis, A. (2024, June 19). *Game Theory: The Mathematics of Strategy and Decision-Making - COMAP.* https://www.comap.com/blog/item/game-theory-the-mathematics-of-strategy-and-decision-making

Applying game theory to Negotiations & Decision-Making | BIO. (n.d.). https://www.bio.org/courses/applying-game-theory-negotiations-decision-making

Ben. (2024, September 9). *How to demonstrate strategic thinking for Government career.* Public Service Careers Australia. https://pscareers.com.au/how-to-demonstrate-strategic-thinking-for-government-career/

Game Theory: Key terms – Economics for everyone. (n.d.). https://econ4everyone.uchicago.edu/game-theory-key-terms/#:~:text=Game%20theory%3A%20The%20study%20of,to%20the%20strategies%20of%20others.

Heiets, I., Oleshko, T., & Leshchinsky, O. (2023). Application of game theory to business strategy. In *IntechOpen eBooks.* https://doi.org/10.5772/intechopen.111790

Kelly, A. (2003). *Decision making using game theory.* https://doi.org/10.1017/cbo9780511609992

London School of Economics and Political Science. (2020, February 20). *Game theory and politics.* https://www2.lse.ac.uk/Events/2020/02/20200220t1830vSZT/game-theory

Ltd, S. (n.d.). *Applying Game Theory in Negotiation | Scotwork Global.* Scotwork. https://www.scotwork.com/negotiation-insights/applying-game-theory-in-negotiation/#:~:text=Strategic%20insight%3A%20Game%20theory%20provides,that%20align%20with%20their%20goals.

Markaki, E., & Chadjipadelis, T. (2023). The use of the Game Theory Context in the strategic political marketing design. The case of the USA elections. *International Conference on Business and Economics - Hellenic Open University, 2*(1). https://doi.org/10.12681/icbe-hou.5352

Mathematical foundations of game theory. (2019, November 19). TSE. https://www.tse-fr.eu/books/mathematical-foundations-game-theory?lang=en

Muñoz, E. Á. (2023, December 29). The convergence of AI and game Theory: Revolutionizing strategic Decision-Making. *Medium.* https://medium.com/@enriqueavila.finance/the-convergence-of-ai-and-game-theory-revolutionizing-strategic-decision-making-91d47695aeda

Parker, E. (2024, January 20). How emotions Influence your Decision-Making - New trader U. *New Trader U.*

https://www.newtraderu.com/2024/01/19/how-emotions-influence-your-decision-making/

The Role of Game Theory in Love/Relationships : Networks Course blog for INFO 2040/CS 2850/Econ 2040/SOC 2090. (2021, September 21). https://blogs.cornell.edu/info2040/2021/09/21/the-role-of-game-theory-in-love-relationships/

The Role of Game Theory in Politics and Society : Networks Course blog for INFO 2040/CS 2850/Econ 2040/SOC 2090. (2019, September 16). https://blogs.cornell.edu/info2040/2019/09/16/the-role-of-game-theory-in-politics-and-society/

Using game theory in business negotiations. (2023, August 28). Elegant Success Shop. https://www.elegant-success.com/blogs/news/using-game-theory-in-business-negotiations

Using game theory to claim advantage in negotiations. (2023, June 21). Kogan Page. https://www.koganpage.com/logistics-supplychain-operations/using-game-theory-to-claim-advantage-in-negotiations

WanderingWriter. (2023, March 21). The Psychology of Game Theory: How Emotions influence Rational Decision-Making. *Medium.* https://medium.com/@bmcgavig/the-psychology-of-game-theory-how-emotions-influence-rational-decision-making-d27e21f93053#:~:text=Emotions%20in%20Game%20Theory&text=Emotions%20play%20a%20significant,the%20outcomes%20of%20the%20game.

Wright, R. (2023, February 7). *Navigating Communication through the Lens of Game Theory: What Works and What Doesn’t.* Communication

Generation. https://www.communication-generation.com/game-theory-in-relationships-a-conversation/

Made in the USA
Middletown, DE
14 April 2025